Yaya Malo

Availability and accessibility to drinking water

Yaya Malo

Availability and accessibility to drinking water

hygiene and sanitation in the southwest region (Burkina Faso)

ScienciaScripts

Imprint

Any brand names and product names mentioned in this book are subject to trademark, brand or patent protection and are trademarks or registered trademarks of their respective holders. The use of brand names, product names, common names, trade names, product descriptions etc. even without a particular marking in this work is in no way to be construed to mean that such names may be regarded as unrestricted in respect of trademark and brand protection legislation and could thus be used by anyone.

Cover image: www.ingimage.com

This book is a translation from the original published under ISBN 978-3-639-50646-4.

Publisher:
Sciencia Scripts
is a trademark of
Dodo Books Indian Ocean Ltd. and OmniScriptum S.R.L publishing group

120 High Road, East Finchley, London, N2 9ED, United Kingdom
Str. Armeneasca 28/1, office 1, Chisinau MD-2012, Republic of Moldova, Europe
Printed at: see last page
ISBN: 978-620-5-83668-2

Contents

To my late father, who left us in August 2010, this work is my gratitude to your blessings and to your fatherly love that you gave me.

To my mother for your love.

To my brothers and sisters, for the moral and financial support you have always given me.

To my friends, for the encouragement and moral support they have always given me.

ACKNOWLEDGEMENTS

We have a great feeling that we have been able to complete this study. It is the contribution of several people whose names we cannot mention.

First of all, we would like to thank the teaching staff of the Geography Department of the University of Koudougou.

We also express our sincere thanks to **Prof. Dapola E. C. DA** for his availability to supervise us and for his advice.

We express our deep gratitude to the USAID WA-WASH Programme for funding this study, where we carried out an internship for twelve months. We are also grateful to the Regional Director of the programme, **Dr Lakhdar BOUKHERROU**. It is our duty to be grateful to them.

We would like to thank Mr. **Jean SIMPORE,** thanks to whom we have consolidated our knowledge of GIS, in particular Arc-View software.

We thank all the staff of the USAID WA-WASH Programme with whom we shared pleasant moments.

We do not forget our fellow trainees with whom we shared a month in the field. They are **Asmaou K. KABORE, Ismaila ILBOUDO** and **Romaric KOUGUHINDIGA**. We simply say thank you.

We thank the Laboratory of Remote Sensing and GIS, University of Ouagadougou, and the students of the said laboratory, who were always present to answer some of our concerns.

We would also like to thank the authorities of the South West region who facilitated our access to the region. They are the president of the South-West Regional Council **Mathias HIEN**, the president of the AMBF of the South-West **Dari SOME** and the different officials of the town halls visited during this study. Finally, we would like to thank the population of the South-West for their collaboration.

SUMMARY

Precarious access to drinking water and sanitation is a hindrance to the development of a country both in economic and human terms. Third world countries in particular and those in Sub-Saharan Africa (SSA) are confronted with problems of drinking water, hygiene and sanitation. According to the 2006 United Nations Development Programme (UNDP) report, lack of water and sanitation leads to 1.8 million child deaths per year in sub-Saharan Africa. Burkina Faso, like other countries in SSA, faces serious problems in this area, both in urban and rural areas. In the specific case of the rural world, this problem is accentuated by the notorious insufficiency of drinking water, hygiene, sanitation and awareness infrastructures. There is therefore a lack and unequal distribution of drinking water, sanitation and hygiene infrastructure in the South West region. The present study aims to highlight the factors that explain the problem of access to drinking water, hygiene and sanitation in this part of the country.

A qualitative approach with quantitative aspects has ёк usedёe. To do this, we did a literature review, ёlaborё and assembled collection tools. We then went to the South West to collect the donations.

And the results show that the efforts already made by the Burkinabё state and development partners have not managed to significantly solve the problem in the South West. She noted that there is an uneven distribution and lack of infrastructure in the region. She also noted that the lack of access to clean water, hygiene and sanitation has consequences for the sank of the region's populations and this is noticeable at the level of the sank centres.

Concepts: Boreholes, Wells; Drinking water; Availability; Accessibilik; Sanitation; Hygkne; GIS.

GENERAL INTRODUCTION

Inadequate access to safe drinking water and sanitation hinders a country's development in both human and economic terms. The understanding of the close link between reliable water supply, adequate sanitation and poverty alleviation has been instrumental in improving access to safe drinking water and sanitation in Sub-Saharan Africa. At the 2013 National Public-Private Partnership Conference, it was noted that *"over the past two decades, more than 230 million people have gained access to safe water and 110 million to sanitation services. But due to Sub-Saharan Africa's high population growth and rapid urbanisation, the efforts made are actually resulting in a small increase in drinking water and adequate sanitation coverage of 11% and 4% respectively over the past two decades. Worse, piped water coverage in urban areas even decreased between 1990 and 2008, from 43% to 35%. Similarly, drinking water coverage in rural areas remains stubbornly low, at around 40-50% during the same period"* (Burkina Faso Ministry of Water, Hydraulic Amenities and Sanitation/DGRE, June 2013; Page 3).

Burkina Faso, being a landlocked country in the Sahelian strip, owes its salvation, in terms of water, to rainfall. Not having a major river as is the case in Niger and Mali, which are crossed by the Niger River, the country sees its water flowing towards neighbouring countries. In addition, the geological nature of the crystalline rocks does not allow for thick, continuous and productive aquifers in some parts of the country, notably in the Birimian, in the South West region. As a result, borehole flows are generally low (2 m^3 /h on average). The Integrated Water Resources Management (IWRM) in 2001 estimated the total water needs in Burkina at about 2,500 million m^3 per year and the consuming demand is estimated at 505 million m^3 /year of which 21% is for domestic consumption. The 2012 Blue Book report shows Burkina Faso to be in a situation of water stress. Indeed, it shows that the average water resources theoretically available and mobilisable are estimated at 850 m^3 per year and per inhabitant, while the shortage threshold is generally estimated at 1000 m^3 (BRUNO V. and MARINA R., 2012).

The situation of access to sanitation is also complex. In 2008, the Office National de l'Eau et de l'Assainissement (ONEA) showed that in urban areas, the sanitation access rate was 17%. In rural areas, surveys by the National Drinking Water Supply and Sanitation Programme (PN-AEPA, 2005) show that the coverage rate is around 1% if only improved technology facilities are considered for the Millennium Development Goals (MDGs). However, the coverage is estimated at 10% considering that a part of the existing traditional latrines, whose rate is estimated at 20% of households, meet the criteria of sëcuritë, effective use and përennitë (Ministry of Agriculture, Hydraulics and Halieutic Resources, 2005).

These very alarming figures are even more so in the South West region. Indeed, access to clean water, hygiene and sanitation is a major problem. Throughout the region, people resort to practices that are not recommended. These include the use of water from non-potable sources for domestic purposes, defecation in the wild and the lack of hygiene facilities. It is with this in mind that we are addressing the issue of drinking water, hygiene and sanitation using the GIS tool. This choice embodies the aspirations of the USAID WA-WASH Programme. Indeed, it is in line with its programme of activities. To do this, the framework of our work is articulated as follows:

- a first part consisting of the theoretical framework and the presentation of the study area ;
- a second part which presents the results of our fieldwork.

PART ONE

THE THEORETICAL FRAMEWORK AND PRESENTATION OF THE
STUDY AREA

CHAPTER 1: THE THEORETICAL FRAMEWORK

In order to carry out this study, it was important to place it in a regional context and to know what work already existed, where it had been carried out and by whom. It was also important to know the problem, to set objectives for solutions, and to formulate hypotheses. In this way, a framework for the field was developed.

1.1. THE SUPPORTING CONTEXT

1.1.1. The problem

For most developed countries, the problems of drinking water, hygiene and sanitation seem to be old memories. In Africa, and more particularly in Burkina Faso, this problem remains. Water, which is at the heart of human activities, whether for domestic needs (drinking, personal hygiene, washing up, laundry) or for economic activities (agriculture, livestock, industry), is not always available.

Since the droughts of the 1970s, water programmes in Burkina Faso have mainly taken into account the difficulties of water supply, while minimising the qualitative aspects. Water is the source of many diseases in third world countries. According to the 2006 report of the United Nations Development Programme (UNDP), lack of water and sanitation causes 1.8 million child deaths per year from diarrhoea in sub-Saharan Africa. In Burkina Faso, waterborne diseases are one of the leading causes of morbidity and hospitalisation.

This is why today, the national water policy accompanies the health policy, through the search for sanitation and hygiene, especially food hygiene. This policy seeks above all to promote drinking water. It has been translated by the vote of the law n°002- 2001/AN of February 8, 2001 which makes water management a national priority.

In order to fight poverty, the United Nations set goals in 2000. One of these goals is the sustainability of ecological resources.

With this in mind, in August 2002, at the Sustainable Development Summit in Johannesburg, South Africa, the goal was set to reduce by half the proportion of the population without access to safe drinking water and sanitation.

Prior to the MDGs, Burkina Faso adopted a National Water Policy in 1998. The overall objective of this policy was to contribute to sustainable development by providing appropriate solutions to water-related problems, so that water would not be a limiting factor in sustainable development. However, in order to achieve the MDGs, Burkina Faso has developed and adopted the National Drinking Water Supply and Sanitation Programme (PN-AEPA). As such, it is the instrument through which Burkina Faso, in line with its Strategic Framework for Accelerated Growth and Sustainable Development (SCADD), aims to achieve the MDGs in the WATSAN sector (Ministry of Agriculture, Hydraulics and Halieutic Resources, 2008).

The NP-WATSAN has clear and quantifiable targets: provision of adequate access to drinking water to 80% and sanitation to 54% of the population in rural areas by 2015; access to drinking water to 87% of the population in urban areas and to sanitation to 57%. For the latter, the National Sanitation Policy and Strategy (PSNA) adopted in 2007, takes a broader view of sanitation than the one used in the NP-WSS sector. Sanitation is therefore divided into four sub-sectors: solid waste; stormwater; liquid waste and gaseous waste (DGAEUE, 2011).

Despite these initiatives, there are concerns about access to safe water, hygiene and sanitation. At the current level, will the MDGs in this area be met? This concern is compounded when one looks at the situation in the South West region, despite its high water potential (acceptable rainfall compared to the rest of the country). In the annual report of the PN-AEPA (2010), the rate of access to drinking water is decreasing in four regions of Burkina Faso, including the South-West region (Ministry of Agriculture, Hydraulics and Halieutic Resources, 2010). For sanitation, in 2008, government statistics and those of the Joint Monitoring Program (JMP) show a national rate of around 10% (BRUNO V. and MARINA R., 2012).

Water needs are determined beyond an incompressible threshold, by habits and uses influenced by behaviour, education and other cultural factors (PROST A., 1996). Water has always conditioned the settlement of people in a locality. Villages, neighbourhoods and towns always have their own water supply system that can probably be improved. Before intervening in this system, it is important to carefully analyse what already exists.

The availability of drinking water is part of a country's well-being and development indicator. The use of pesticides and chemical fertilisers in agriculture pollutes surface and ground water. In addition, gold mining in the Birimian-dominated South West region pollutes rivers with chemicals (cyanide, mercury). Human excreta are not to be outdone by a population that mostly relieves itself in the wild. Human and animal excreta produce nitrate which is washed into the rivers. The rivers, being the lowest points, are at the same time a "bin" because all the waste ends up there. Access to clean water, a healthy environment and adequate hygiene is therefore a concern for the people of the South West. In order to find solutions to this problem, we have asked ourselves a number of questions.

1.1.2. The research questions

The general question that emerges is: What are the factors influencing the non-availability and non-accessibility of drinking water, hygiene and sanitation in the South West region? This is broken down into three sub-questions.

a) What is the availability of drinking water in the South West region?

b) What is the level of access to drinking water in the South West region?

c) What is the level of access to hygiene and sanitation in the South West?

To answer all these questions, hypotheses have been formulated.

1.1.2. Research hypotheses

The general hypothesis is that the inadequacy and uneven distribution of drinking water points and hygiene and sanitation infrastructure are constraints to improving the living conditions of the people of the South West Region. This hypothesis is supported by three other so-called secondary hypotheses:

a) the uneven distribution of water points complicates the availability of water for the populations of the south-west region;

b) People in the South West have difficult access to water;

c) There is a low level of access to hygiene and sanitation in the region.

1.1.3. Research objectives

The general objective of this study is to identify the factors that contribute to the low availability and accessibility of drinking water, hygiene and sanitation facilities in the South West region, as well as water-borne diseases. Three specific objectives are derived from this:

a) to take stock of the water supply points for the population in the South West region;

b) dëfmir the level of accessibility to drinking water in the region;

c) determine the level of hygiene and sanitation infrastructure in the region.

c.2. RESEARCH METHODOLOGY

The research methodology implemented is a qualitative approach, with quantitative aspects. It consists of a literature review, the development of survey tools, data collection, processing and analysis.

c.2.1. The literature review

Before starting this study, we looked for previous work in the WASH sector. We were able to find work at various scales around the world.

On a global scale, several initiatives are being taken and work has been done to provide solutions to the problems of drinking water, hygiene and sanitation. This WASH problem is not new. For PROST A. in Population et Environnement dans les Pays du Sud, (1996) page 231 *"the unhealthy character of wetlands for humans has been a fact of medicine and even of popular wisdom since antiquity until the 19th century. The idea of this is argued from the 5th century BC in the treatise on areas, waters and places. A certain relationship had been established between these unhealthy environments and diarrhoea or rather intestinal diseases as a whole. But water had never been identified as the vehicle of disease, nor was its quality considered to be of medical importance. It was not until 1854 that John SNOW, a London GP, had the Broad Street public fountain condemned in the midst of a cholera epidemic. In a few days, the number of new cases collapsed among the families who drank from it and his act imposed the notion of contamination by water.* And more recently, a study of 20 urban concentrations in Brazil estimated that drinking water supply was responsible for one-fifth of the reduction in infant mortality between 1970 and 1976 in the areas served (MERRICK, 1985).

Today, more than ever, the issue of water and sanitation is at the heart of major debates, especially in the context of climate change and pollution. The United Nations (UN) and the UN agency specialised in heritage protection (UNESCO) organised the International Year of Freshwater in 2003. According to the results of their surveys, 1 in 6 of the world's inhabitants do not have access to a proper supply of drinking water. In addition, 2 out of 5 people do not have access to sanitation, i.e. wastewater disposal.

Major international meetings have been organised (Mar Del Plata, 1976; Dublin and Rio de Janeiro, 1992; The Hague, 2000; Kyoto, 2003), with the aim of alerting mankind to the ecological risks and promoting a healthy life in harmony with nature. International organisations and programmes have emerged to help countries improve their WASH situation. These include: Global Water Partnership or (World Water Council), International Water Management Institute or International Hydrological Programme. At the recent World Water Forum in Marseille, in March 2012, it emerged that: "every euro invested in water management and especially sanitation saves 5 to 6 euros in public expenditure". His words attest to somewhat similar analyses by WALSH and WARREN at the Rockefeller Foundation (1979), which show that the cost of a "quick death" could be as high as $3,000 for a programme focusing on nutrition and $4,300 for a water and sanitation programme.

At the African and especially sub-regional level, the end of the International Drinking Water Supply and Sanitation Decade (IDWSSD) has necessitated the creation of a regional structure that develops and popularises WATSAN techniques and approaches adapted to local contexts. It is in this context that the Centre Rëgional pour l'Eau Potable et l'Assainissement (CREPA) renamed Water and Sanitation for Africa (EAA), in 2012, was cтёë in 1988 under the impetus of the UNDP Water and Sanitation Programme and the World Bank, the Swiss Coopëration and the Ecole Polytechnique Fëdërale de Lausanne. Its mission is to promote access to safe drinking water and basic sanitation services. It is primarily aimed at low-income populations living in rural, peri-urban and urban areas. It specialises in training and research for the extension of appropriate technologies in the WASH sector. The preferred strategy is a participatory, operational and financial approach focused on the needs of the poorest communities. The EAA has a spatial database on the implementation of Ecosan latrines, (WINKLER S., 2005). According to its former General Director, TANDIA C. T. (INFO

CREPA, 2004, n°43) *"the financing of the WASH sector has always been included in the State budget as a public investment and until now, it has been ensured in large part by external resources. This leads to a strong dependence which has a huge influence on the taking of initiatives or their implementation, or the sustainability of actions".*

At the national level, policies exist for WASH. Table 1 shows the history of the different policies adopted and major changes in the field of WASH in Burkina Faso.

Table 1: Evolution of WASH policies in Burkina from 1970 to 2009.

Years	Events
1970	Separation of water and electricity management through the creation of the Societe Nationale des Eaux (SNE), a mixed economy company which operates in 7 urban centres while the State directly supervises rural water supply.
1976-1978	First water policy and nationalisation of the SNE, transformed into a National Water Office operating in 44 urban centres.
1985	Transformation of ONE into Office National de l'Eau et de l'Assainissement (ONEA) and creation of a sanitation fee on the water bill.
1994	ONEA becomes a state-owned company. The sanitation fee is mobilised to finance the Ouagadougou Strategic Sanitation Plan (PSAO).
1998	The adopted National Water Policy introduces integrated water resources management in Burkina Faso, which is a first in West Africa.
1998	Adoption of the "Textes d'Orientation de la Decentralisation" or TOD.
1996-2000	ONEA improves its technical and financial performance.
2000	Reform of the management system of hydraulic infrastructures for drinking water supply in rural and semi-urban areas.
2001	Adoption of the Water Law.
2002	Creation of the Ministry of Agriculture, Hydraulics and Halieutic Resources (MAHRH), now called the Ministry of Water, Hydraulic Amenities and Sanitation.
2003	Adoption of the Action Plan for Integrated Water Resources Management (PAGIRE).
2006	Adoption of the National Drinking Water Supply and Sanitation Programme (PN- AEPA) and development of its implementation tools.
2007	National Sanitation Policy and Strategy (NSPS)
2008	Institutional separation of water and sanitation management in rural areas with the creation of the Direction Generale de l'Assainissement des Eaux Useseses et Excretas (DGAEUE) alongside the Direction Generale des Ressources en Eau (DGRE).
2009	Decree on the transfer of water and sanitation competences to the municipalities.

Source: Blue Book Country Report, 2012.

Despite these different policies, the issue of drinking water, hygiene and sanitation is still an ongoing problem.

The publication by SAVADOGO A. N. (2012), shows the limits of current policies, after having made a synthesis of the different drinking water policies. He questions certain policies and proposes to move from supplying drinking water to villages through isolated boreholes equipped with human powered pumps, to distributing water to these villages through networks connected to water towers of appropriate capacity, which would be supplied by boreholes drilled on mega-fractures and their nodes. We derive the notion of wellfields from this because the required number of boreholes will all be drilled on a single large fault that will be intersected by secondary fractures. He geo-referenced the wellfields. They are defined as a space containing distension faults, at best, injected with quartz or pegmatite veins, or fault nodes capable of supplying the wells that are installed there with large flows (over 20 m^3 /hours).

NAKOLENDOUSSE S., (1991) shows that the poor distribution of rainfall in time and space, the geographical and geological conditions, make surface water insufficient in Burkina Faso, not to mention its doubtful potability for human consumption. The use of groundwater is essential, but due to the complexity of the discontinuous aquifers in the basement terrain that make up 80% of Burkina Faso's territory, this water is difficult to access. It also shows that work on groundwater exploration and exploitation in the basement regions has made it possible to clarify their mode of deposition. It is known that the water tables in these regions are small and that they present the aspect of a three-dimensional network of faults and fractures of variable extension, characteristic of this type of geological substratum. Thus, methods for locating the water table have made it possible to improve the location of groundwater catchment structures; but the flow rates of these structures vary enormously from 0 to 50 m /h.^{3}

HUGOT G., (2002) in his paper entitled Lost Gondwana shows that the Birimian basement, due to the rigidity of their rocks, fractured abundantly in response to the different tensions that were exerted on them during the Precambrian and Paleozoic orogenies. These phenomena would be decisive in the Birimian of the region. Indeed, they would explain the faults of the region as well as the model which is much more undulating and presents more incised talwegs. The glacis on the granite-gneissic base are long and rectilinear slopes, with an iron shell. The interfluves are installed on much thicker slopes and the base is no longer outcropping. The lowlands are less wide and more incised, but they form a dense network.

OUANGRE J. V. T., (2009) finds that exposure to water-related health risks differs across countries and environments. Without clean water supply, basic sanitation and good water resource management, the goal of reducing water-related diseases will not be achieved. However, the provision of water points, the organisation of water distribution, the management of pollution sources (waste and excreta) are the basis of water hazards in a given environment. The effectiveness of source water protection measures and the improvement of water quality depend greatly on a society's perception of the role of water in the emergence and spread of disease. Knowledge of community perceptions of the role of water in disease phenomena is an essential component of all water-related disease control strategies. Indeed, simple actions, well understood by the population, can be life-saving.

At the commune level, sectoral WATSAN Community Development Plans (CDPs) have been produced. There are also regional and provincial monographs that take into account drinking water, hygiene and sanitation issues. Very detailed work on the issue of drinking water, hygiene and sanitation is rare in the region, apart from that of certain NGOs.

c.2.2. Definition of concepts

Accessibility: is the character of what is accessible according to the Larousse illustre, (1992). Access to drinking water refers to the problem of means, time, in short the organisation that will bring water to consumers (wikipedia.org, June 2012). Access to drinking water in the monitoring and evaluation manual of the PN-AEPA (MAHRH, 2007, annex 3, page 1) is the percentage of the population that can access, under satisfactory conditions, a sufficient supply of drinking water, at home or at a reasonable distance from it. For our purposes, access to drinking water is defined as a household that does not make sufficient effort to access it.

Availability: this assumes the existence of a quantity of water above the scarcity threshold estimated at the international level at 1,000 m^3 of water per capita per year. A quantity below this threshold corresponds to water stress. Permanence indicates a continuous and sustainable supply of the commodity. The distance expresses the norm for access to the network, i.e. to be located at least 50m from distribution pipes and less than 500m from standpipes or common water points. However, according to ONEA, "beyond 50m, an extension may be offered if there are a number of applicants (at least 5) over a distance of 200m for example (www.oneabf.com).

The Geographic Information System (GIS) is a tool to assist in decision making. It originated in Canada in the 1960s. It is defined by the Sociëtë Franchise de Photographie et de Teledetection, (1989) as a "computer system allowing, from various sources, to gather and organise, manage, analyse and combine, elaborate and present geographically localised information, contributing in particular to spatial management". In other words, GIS enables the processing, management, analysis, integration and modelling of geographical data, as well as the processes that transform the territory. As such, it is necessary to manage integrated, multi-dimensional information representing complex environments, in addition to being able to model development scenarios. GIS contributes to the improvement of decision making in terms of access to drinking water and sanitation, for the benefit of the various State services, managers and users.

Hygiene: according to the Encarta dictionary, it is a set of individual or collective practices aiming at the preservation and improvement of health. And according to the Larousse (1992), it is the part of medicine that studies the individual or collective means, principles and practices that aim to preserve or promote health. For us, hygiene is the care of our body, food and objects to keep them clean.

Sanitation: is an approach to improving the overall health status of the environment in its various components (Wikipedia.org, June 2012). It is defined by WHO as: "action aimed at the improvement of all human conditions that affect or are likely to affect adversely physical, mental or social well-being". Literally, it is "the action of sanitising", and refers to all techniques and methods aimed at treating wastewater and man-made waste. Sanitation is strongly linked to public health because of the many diseases that result from an unhealthy environment. Proximity to wastewater can lead to diseases of fecal-oral transmission (diarrhoea, typhoid, hepatitis, cholera), or related to poor basic sanitation and especially to defective or non-existent latrines: bilharzia, nematodes or other worms. For us, sanitation is about making our living environment clean.

Drinking water: The World Health Organisation (WHO) has established a number of guidelines regarding the quality required for water to be considered safe to drink. These guidelines are the international references that guarantee safe and therefore drinkable water. The latest guidelines are those issued by the WHO in Geneva in 1993. Drinking water must be free of suspended solids, micro-organisms and toxic substances. The recommendations for mineral concentrations vary from country to country, but for most minerals a maximum concentration is recommended to ensure balanced and pleasant drinking water. (http://www.lenntech.fr/applications/potable/normes/normes-eau-potable, 20/11/2013).

The Council of the European Union issued a directive 98/83/EC on the quality of water required for human consumption. It was implemented on 3 November 1998. It was drawn up by taking the parameters of the 1980 Drinking Water Directive and incorporating where necessary the latest scientific findings on the effects of different substances on humans (WHO guidelines and the World Scientific Committee on Toxicology and Ecotoxicology). The thresholds below are the main changes made:

- lead: the accepted dose has been reduced from 50 gg/l to 10 gg/l, with a 15-year tolerance to allow manufacturers to replace the lead on their production lines.

- pesticides: the values for individual substances as well as for total pesticides have been maintained (0.1 gg/l / 0.5 gg/l), and stricter values have been introduced for some specific pesticides (0.03 gg/l).

- copper: the value has been reduced from 3 to 2 mg/L.

- some new limits have been introduced for elements such as trihalomethanes, trichloroethene, tetrachloroethene, bromate, acrylamide etc. We therefore say that water is potable when it meets a number of characteristics that make it fit for human consumption.

The borehole: the borehole is defined by BENAMOUR A., (1981) *in* GALBANI S. R., (2011) as "a very narrow well (15 to 20 cm in diameter) collecting clear water at a depth of 50 m or more. Its walls are lined with a metal casing". Boreholes are in fact structures equipped with a pumping system that brings water up from the depths. Stainless steel pumps are most often used. The pumping is done by man, hence the term human powered pump (HPP).

The well: modern wells are large or small diameter structures designed to retain water from the water table. They are called modern because they are made of reinforced concrete over their entire depth. These wells are equipped with a bottom slab and a concrete coping, with an average height of 0.80 m and an internal diameter of 1.8 m for large diameter wells and between 0.8 and 1 m for small diameter wells (GALBANI S. R. P., 13 2011). As for the traditional well, we define it in this study as a vertical hole, most often circular and with non-magoon walls, without a coping, generally dug in the ground to reach the water table. It is made by the populations with traditional know-how.

1.2.3. The study area

The region is located in the southwest of the country between 9°20' and 10°50' north latitude and 2°25' and 5°25' west longitude. It has a surface area of 16,318 km^2 , i.e. 6% of the national territory and an estimated population of 620,767 inhabitants in 2006, i.e. 4.42% (RGPH, 2006) of the population of Burkina Faso, with a density of about 38 inhabitants/km^2 . It is bordered to the east by the Republic of Ghana and the Centre-West region, to the north by the Hauts-Bassins, Boucle du Mouhoun and Centre-West regions, to the west by the Cascades region and to the south by the Republic of Côte d'Ivoire. Two main axes cross the region from west to east.

These are: National Road No. 6 in the north, which connects Bobo to Leo, via Diebougou, and in the south, No. 11, which connects Banfora-Gaoua-Batie. Another road links Diebougou-Gaoua-Kampti-Frontiere Cote d'Ivoire via Galgouli (see map 1).

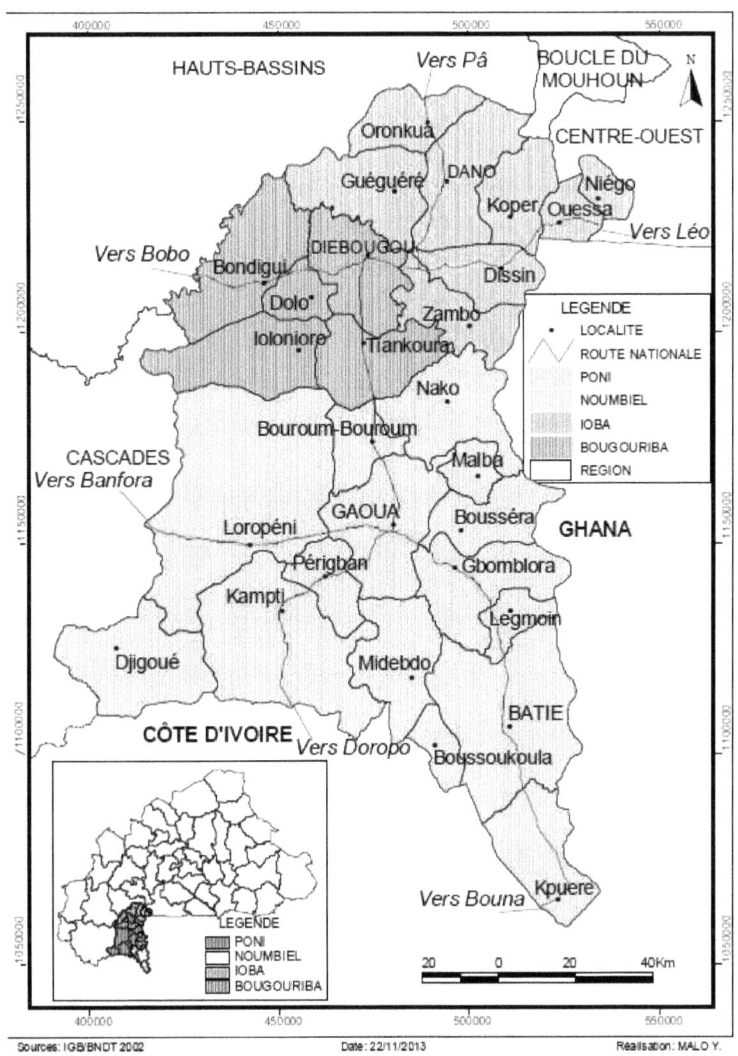

Map 1: The South West region

1.3. THE APPROACH TO DATA COLLECTION

It takes into account sampling and preliminary work including the development of questionnaires and interview guides, the collection of data and the test survey.

> Sampling

In the study area, we selected a sample of 200 households from the population of our target villages. The sample mainly concerned the 40 villages of intervention of PROMACO (in charge of ensuring the distribution of the tablets in the region, with a focus on the 40 villages) in the South-West, i.e. about 3% of the households. The households were distributed among 24 villages, i.e. 60% of the total number of intervention villages, and the remaining 40% were subject to a simple census of water points.

The 24 villages were selected randomly. To do this, a number was assigned to each of the selected villages in each commune. The objective was to reach at least 50% of the villages. A rule of three was used at all stages to determine the number of households to be surveyed per province and per commune. For the number of households to be surveyed per village, we divided the number of households per commune by the number of villages to be surveyed in that commune. The results are presented in Tables 2 and 3.

Table 2: Villages with household surveys and water point surveys

Provinces	Municipalities	Villages	Number of households to be surveyed
Bougouriba	Bondigui	Bondigui	11
		Mougue	10
	Diebougou	Bapla	5
Ioba	Dano	Balembar	17
		Sarba	17
		Tambalan	17
		Tambiri	17
	Gueguere	Dakoula	8
		Moulourou	8
	Koper	Bene	7
		Babora + (Pirkon)	7
	Niego	Bevougane	7
		Dadoune	7
	Ouessa	Ouessa	17
	Zambo	Bontioli	13
		Zambo	13
Noumbiel	Batie	Koudjo	5
		Poni-kienkire	5
	Midebdo	Kalambouro	2
		Kpanhila	1
	Legmoin	Dakpolo-gompar	2
Poni	Boussera	Gbon	1
		Koubeo-djoulo	2
		Kpara	2

Source: PROMACO

Table 3: Villages with a simple water point census.

Provinces	Municipalities	Villages
Ioba	Dano	Complan
		Dayere
		Yo
		Yobogane

	Gueguere	Bile
		Katogue
	Koper	Dalgane
	Niego	T amagane
	Ouessa	Dianle
	Zambo	Djikpologo
		Kpankpire
Noumbiel	Batie	Tamipar
	Midebdo	Kpanhila
		Torkouora
	Legmoin	Douotaon
Poni	Boussera	Kouleho
		Tankolon

Source: PROMACO

> The variables and indicators of the study

The variables

Independent variables: inadequacy of thc systcm

The dependent variables: bad behaviour

The indicators

Inadequacy of the system

- lack of drinking water infrastructure

- lack of awareness programme

- insufficient supply of drinking water

- lack of water point management structures

- sources of drinking water supply ëloignëes

- drinking water sources near rubbish dumps

- lack of individual and public latrines

- lack of bins

Bad behaviour

- lends water to dubious water sources

- dëfëque in nature

- living in an unhygienic environment

- does not treat drinking water

- does not clean the water drawing equipment

- does not take hygienic precautions when transporting water

- do not clean hands before drinking water collection

- do not clean hands before and after meals

- do not clean hands with soap after using the toilet

- throws the rubbish in the yard

- throws the water into the courtyard

> Preparation for data collection

Before going to the field, we gathered the necessary material for the data collection. These

were: a GPS (GARMIN etrex 30), a camera, a calibrated rope, a household questionnaire and

19

another for water point managers and; interview guides addressed to opinion leaders and health workers. A field trip was made to Tanghin-Dassouri, during which we tested our household and water point manager survey forms as well as our different interview guides for health workers and opinion leaders. The other data collection tools (GPS and camera) were also tested.

1.3.1. Data collection in the field

It is carried out in all provinces of the South West and lasted one month (10 December 2012 to 10 January 2013). It consisted of GPS surveys of water supply points, households, latrines and sociological surveys. Two teams were formed for this data collection. It was supervised by the USAID WA-WASH GIS and Food Security Officer.

Collecting the coordinates of water supply points.

It involved all water supplies. Rubbish dumps and latrines were included in the households because of their proximity to them. These surveys provided us with a database that enabled us to produce the various maps.

> The sociological survey

It provided the opportunity to obtain the population's opinion on certain issues concerning drinking water, sanitation and hygiene. The information collected from households and water point managers was used for the statistics. As for the different interview guides, they allowed us to have another point of view and explanations to this problem.

1.3.2. Data processing

GPS data

The GPS data were transferred to the computer using OziExplorer software and processed with Arc-GIS (version 10), Google Earth and Arc-View 3.2a. These GIS softwares allowed the production of thematic maps of the region. The surveys carried out allowed us to produce maps of the distribution of water points by type and those of dwellings.

> Processing the survey forms

After collecting the data from the households, we first proceeded to a manual tabulation. The forms were then sorted by province, then by commune and finally by village. Computer processing was carried out using Microsoft Office 2010 software (Excel and Word).

1.3.3. The issue of the study

The purpose of this study was to provide decision-makers in the region and the various stakeholders with a tool that will enable them to act effectively in the field of drinking water, hygiene and sanitation. It already serves as a reference database for the USAID WA-WASH Programme, which aims to enable this population to clean up their drinking water at home, as well as their living environment. In fact, the programme wants to provide the population with water purification tablets (aquatab) to purify their drinking water.

CHAPTER 2: THE PHYSICAL ENVIRONMENT OF THE SOUTHWEST REGION

The region has slightly different physical features in terms of climate, vegetation, fauna, hydrography, relief and flora. These elements are very important factors in the formation and supply of groundwater and in the development of surface water.

2.1. GEOLOGY, GEOMORPHOLOGY AND PEDOLOGY

2.1.1. Geology

2.1.1.1. The Precambrian

It is subdivided into Lower and Middle Precambrian.

The Lower Precambrian or Antebirimian, represented by granite gneisses or migmatites, amphibolites, gabbro and granite, forms the ancient framework of the region. The composite units called craton by ARNOULD M., (1961) are from west to east :

the Mangodara massif, most of which is located in the Banfora region and only its eastern edge appears on the western side of the South West region;

the Bouroum-Bouroum massif, which is the northern extension of the Bouna craton massif;

and the Mouhoun (ex Black Volta) wrinkle, is the much narrower granite-gneissic domain. It is largely located in Ghanaian territory.

The Middle Precambrian or Birimian: represents the non-eroded parts of the intra-cratonic trench fill of volcanic, pyroclastic and sedimentary origin with the following stratigraphic succession from base to top:

basaltic and andesite formations ;

S volcanic and volcano-sedimentary formations combining calc-alkaline dominant lavas, basic and neutral tuffs, shales, varying levels of chemical sediments, detrital sediments.

The Birimian consists of two units located between the granite-gneiss moles. These are :

- The Burkinabe part of this ensemble is referred to as the sërie de Houndë, a locality situated in the north of the South-West region;

- and the eastern unit which is the Gaoua series, (MARCELIN J., 1971). The Birimian series represent the non-eroded part of the intracratonic furrow filling. This filling is of volcanic, pyroclastic and sedimentary origin. The Hounde series is clearly sedimentary dominated, whereas volcanism is more important in the Gaoua series. These formations are folded and metamorphosed. The metamorphism is epizonal and the main folding has a vertical or subvertical axial plane oriented north-south, (MARCELIN J., 1971). It is at the level of the faults of these formations that one can implant the drillings, the AEPS etc.

2.1.1.2. Minerals

According to the explanatory note of the geological map (Gaoua-Batie sheet, 1971), the mineralisation comprises the following minerals, disseminated in veinlets or specks scattered in the altered and fractured diorite

- abundant sub-automorphic pyrite with chalcopyrite and pyrrhotite inclusion;
- molybdenite, in very fine lamellae disseminated in a bluish quartz gangue, in the form of rare independent veinlets, with some sericite, chlorite, epidote and late pyrite. The gangues, i.e. the transparent minerals accompanying the sulphides, are essentially quartz, chlorite and sericite.

In addition to these minerals, there are also deposits of certain minerals:

- The Dienemera deposit, about 20 km west of Gauoa, is located in microdiorites and diorites, quartzites associated with porphyry andesites, all intrusive in the diabases of the Gaoua series. The mineralisation occurs within intensely fractured zones in which the diorite is affected by a meridian-trending diabase. The phenomënes of hydrothermal alteration are clearly visible in the field. The friable diorite is white, mauve or purple in colour. In thin sections, there is an invasive development of chlorite, carbonate, sericite, epidote, ^^xë^. It is a propylitization and saussuritization of feldspars, chloritization of biotite, transformation of hornblende into tremoulite and chlorite;

- The Gongondy deposit, located 8 km south of Gaoua, is of the same type, the mineralisation being linked to another dioritic intrusion located in the extension of the Diënëmëra one. The reserves of copper sulphide are, according to the drillings, about 15 million tons of ore. In contrast to Diënëmëra, phënomënes of cëmentation in a fracture resulted in a small lens of very chalcosine-rich ore, which has been fully depleted.

2.1.1.3.   General characteristics of aquifers.

According to the Direction de l'Inventaire des Ressources Hydrauliques (DIRH, 1993), there are

Hydrogeological ensembles

The South-West region is made up of hydrogeological units defined on the basis of lithological, tectonic, hydrogeological and geomorphological data. The basement is composed of rocks of middle Precambrian age: granites, migmatites, meta-volcanites, volcanic-sedimentary rocks and clayey or clayey-grey meta-sediments. Within the ensemble, two subsets can be distinguished, granite-gneissic and schisto-gresous. The latter is characterised by a high thickness of alteration. The sedimentary ensemble of Upper Precambrian age is also represented by the shale-interbedded greseous rocks. Superficial formations of Quaternary age are largely represented by laterites and alluvium.

The aquifer system

The southwestern region is constituted by the fractured zone, altered zone and laterite aquifers. Generally, the laterite aquifers are in hydraulic contact with those of the underlying fractured rocks; they form a bi-layered system with the altered zone being predominantly capacitive and the fractured medium predominantly conductive. The altered zone can be described as a medium with gap porosity. The underlying fissure medium, on the other hand, is heterogëne and its hydraulic characteristics are determined more by the density and geometry of the cracks. These cracks close progressively with depth. The hydraulic continuity of the hydrogeological systems is not yet clear. However, it is noted that :

the alternating zone with interstitial porosity, can ensure hydraulic continuity;

in fissured environments, continuity depends on the interconnection of fissures. The altered mantle with a thickness of 10 to more than 50 m covers the basement and contains the continuous aquifère which replaces the discontinuous aquifère at the level of the fissured zone with an alteration thickness of 40 m. The results of numerous boreholes show the saturated alteration thickness as the guarantee of the durability of the structure. Almost all of the area's bedrock is covered by 10-30m of alteration and flows vary by $0.5m^3$ /h on average. This is important for the availability of the resource on the one hand and allows the water to mineralise.

2.1.2. Geomorphology

The relief of the region is very uneven with an average altitude of 450 m. However, there are topographical units: vast plains, lowlands, hills, and hillocks. The average altitude of the hills varies between 300 and 500 m (see map 2). There is a succession of hill ranges from Dano to Diebougou. This is what gave the province its name. The main ranges are located south-east of Karangasso, in the Gaoua region, south of Kampti, south of Batie and around Kpuere. The highest point in the region is Mount Nouehe or Koyo, whose bauxite cap is 592 m high. The hydrographic network is moderately dense and hierarchical. The lowlands are slightly incised with large wide valleys. In the west, the bed of the Black Volta (present-day Mouhoun), which is not very deep, slopes gently from 240 m in the north to 220 m in the south (MARCELIN J., 1971). This relief is very important for the realisation of water reservoirs which can, if they are, support the water table.

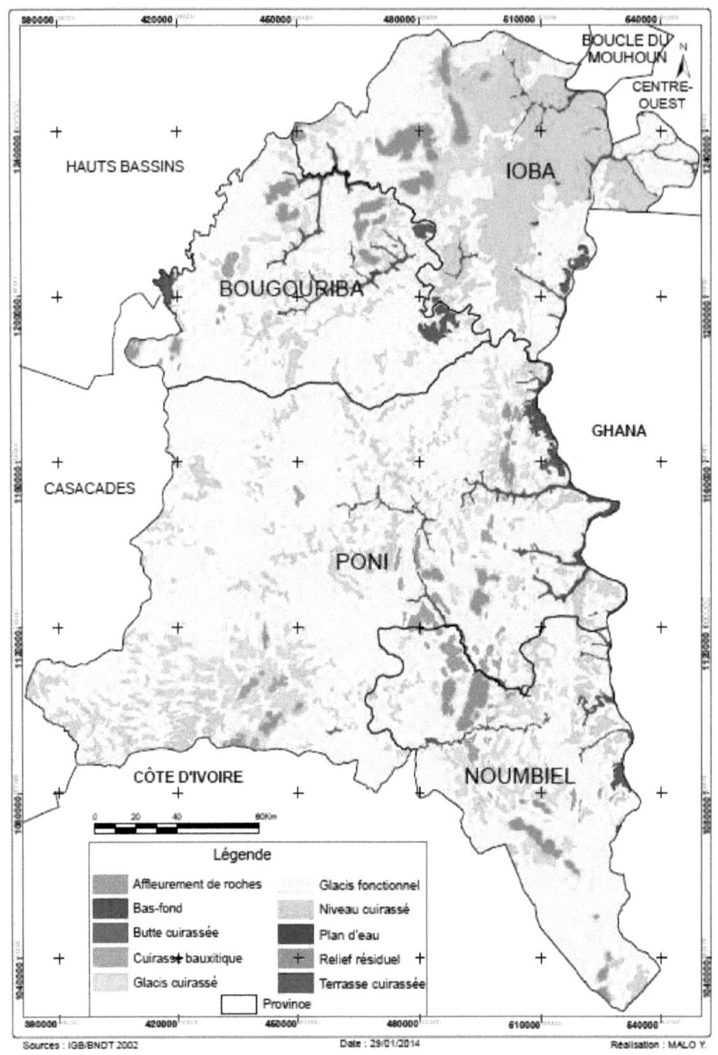

Map 2: Geomorphology of the South West region

2.1.3. Soils

The soil resources of the region are divided into 9 classes by the Bureau National des Sols (BUNASOLS, 1976). This is based on the FAO classification. In the South-West region, 7 soil classes are encountered: soils with sesquioxides and rapidly mineralised organic matter, mull soils, hydromorphic soils, fersiallitic soils, crude mineral soils and vertisols, (Profil des régions au Burkina Faso, 2009). These soils are described as follows by BUNASOLS:

Soils with sesquioxides and rapidly mineralised organic matter are soils with very deep, primitive mineral alteration and generally deep-lying while the surface organic matter undergoes a very rapid evolution. This type of soil is the most common in the region. This soil class includes the various sub-classes of tropical ferruginous soils associated with sandy to sandy-clay soils, with gravel at depth. Soils with sesquioxides and rapidly mineralised organic matter are characterised by their very high iron or manganese oxide and hydroxide content, which gives them a red, ochre or black colour. They have a low natural chemical fertility. Millet, sorghum, groundnuts, mat, cotton and other crops are usually grown on these soils. Their exploitation requires the use of organic and chemical fertilisers (NPK), to obtain good yields.

mull soils belong to the group of eutrophic brown soils. They are associated with volcano-sedimentary formations. Their composition is sandy-clay on the surface and clayey at depth. In the South-West region, this class of soils is mainly found in the areas of Diebougou, Gueguere, Gaoua, Batie and Kampti. These soils are characterised by a humus with high biological activity, a good structure and a complex with high calcium saturation. They represent the best soils in the country and are easy to work. Their thickness is over 1 m. These soils are suitable for a wide range of crops such as mat, cotton, sorghum, sugarcane and fruit.

hydromorphic soils are deep soils (over 100 cm) with poor drainage. This class of soil is suitable for rainfed and irrigated rice and market gardening.

The fersiallitic soils found in the Bati' department are tropical ferruginous soils that are not or only slightly leached and not associated. Because of their low chemical content, their exploitation must be accompanied by the addition of organic fertiliser. They are used to grow cereals (millet, maize, sorghum), groundnuts and cotton.

The raw mineral soils identified in the region belong to the lithosol subgroup. This soil class is located in Ouessa (Loba province). Consisting mainly of ferruginous cuirass on a flat terrain, the raw mineral soils lack a sufficient base for root development. Their agronomic interest is therefore low to nil.

J poorly evolved soils occur as dissolved patches in the region. They have a low chemical water holding capacity due to their coarse texture, limited soil thickness and runoff losses.

Chemical fertility is a function of the geological nature of the bedrock but is generally low. These soils are used for millet and peanut cultivation and can also be used for cattle grazing.

J Vertisols develop mainly on basic rocks and alluvium. In the South-West region, they are located along the rivers to the north and south of Gaoua. They are characterised by a high content of swelling clay, which corresponds to a high capacity for change and the formation of cracks following successive desiccation and wetting of the environment. They are chemically rich but are very difficult to work with traditional tools. The adoption of mechanical tillage allowing deep loosening would give interesting yields on these soils for the production of maize, sorghum and cotton.

These soils (see map 3) are, on the whole, undergoing a strong gradation linked to anthropic actions (extensive production systems, abusive cutting of wood for energy, anarchic occupation of the land). These soils also influence the chemical composition of the water.

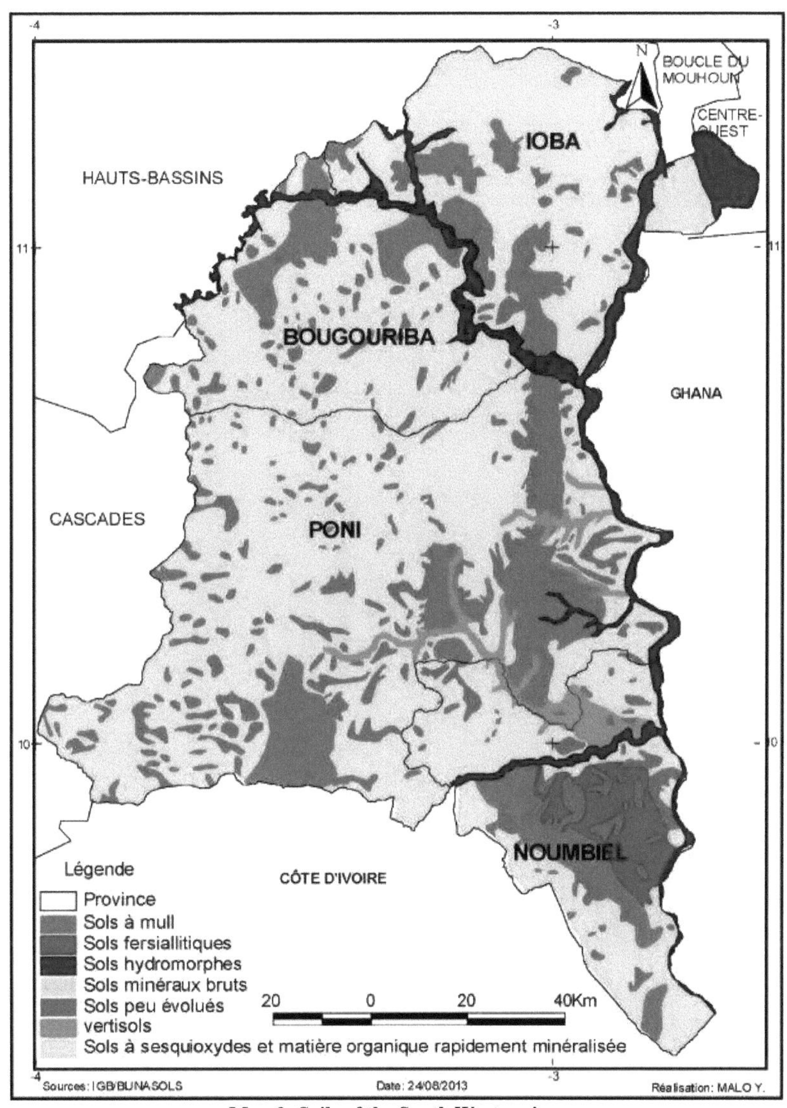

Map 3: Soils of the South West region.

2.2. CLIMATE AND HYDROGRAPHY
2.2.1. The climate

The climate of the South West region is essentially South Sudanese, with two distinct seasons. A dry season that lasts about 5 months (November-March). It is marked by the harmattan, a dry, dusty wind that blows from November to March with mild temperatures of around 27°C. These temperatures generally oscillate between 21°C minimum and 32°C maximum. The rainy season lasts about 7 months (April to October) with maximum rainfall in July and August. The wet period is characterised by the rise of the front in April to the extreme north of the country and its descent towards October. The average annual rainfall is between 900 and 1300 mm. Figure 1 below shows the monthly average rainfall and temperature from 1981 to 2010, i.e. 30 years, for the Gaoua station. The climate is an important factor for the availability of water in a locality.

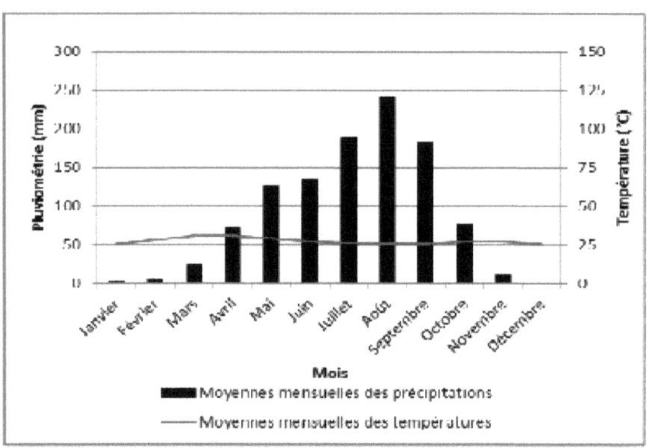

Data source: Direction Generale de la Meteorologie, Ouagadougou (2012).
Figure 1: Rainfall-thermal diagram for Gaoua (1981-2010).

2.2.2. Hydrography

The hydrographic network is represented by two international watersheds, namely the Mouhoun and Comoe basins. The Comoe basin is fed by the Baoue (Keleouoro) and the Iringou (Koba). As for the Mouhoun basin, it is fed by rivers, the main ones being: the Bougouriba, which is made up of the Maoulindara, the Baplakola and the Lougouano, and the Bambassou, which is the union of the Poni and the Kamba, the Pouene and the Koulbi. The Mouhoun and the Bougouriba (on part of its course), are the only permanent rivers. The others are reduced to a string of ponds in the dry season. Thus, the region has sufficient water and important potential dam sites such as the Bougouriba and the Noumbiel (see map 4).

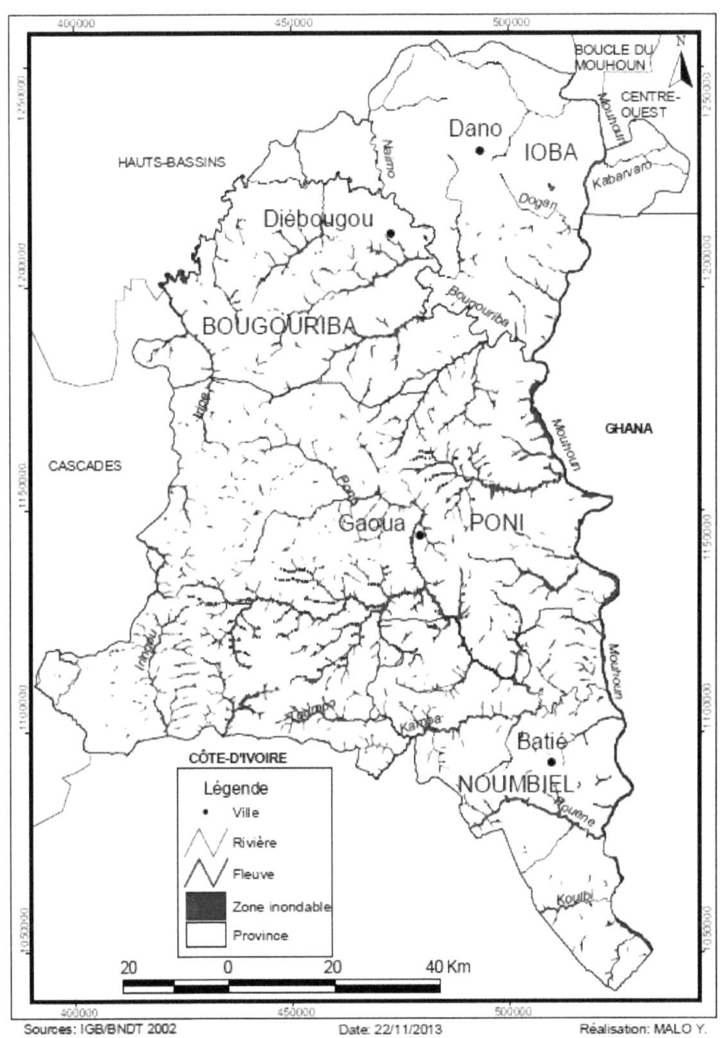

Map 4: The river system of the South West region

2.3. FLORA AND FAUNA

The southwestern region belongs to the Sudanian phytogeographic domain with a savannah vëgëtation. From north to south, there is an evolution from shrub savannah to open forest and gallery forests along the rivers. The shrub savannah covers the provinces of Ioba and Bougouriba for the most part, with a densification from north to south. The Sudanian wooded savannah is found in the communes of Zambo, Ouessa, Niëgo (Ioba province); the communes of Tiankoura and Diëbougou in Bougouriba; the communes of Nako, Malba, and Kampti in Poni. It consists of shrubs and trees forming a clear canopy, allowing light to filter through (Profil des régions au Burkina Faso, 2009).

The pre-Guintentee woodland savannah is found in the plains of Loroptni, Gaoua and Gbomblora in the Poni province and Legmoin in the Noumbiel. Like the Sudanian woodland savannah, its stratum varies between 5 and 12 m and the cover rate is 20-70%. The open forest is present further south in the region, notably at Batit, Midebdo, Kputrt, in Noumbiel and Djigout, in Poni province. It is dense with a woody cover of 70% to 90%. The stratum varies between 20 and 30 m in height. Trees such as *Parkia biglobosa*, *Vitellaria paradoxa*, *Pterocarpus erinaceus*, *Burkea africana*, *Acacia nilotica*, *Combretum nigricans*, *Crossopteryx febrifuga*, *Mitragyna inermis* are found (GUINKO S. et al, 1995).

According to the Ministëre de l'Economie et des Finances (2005), the protected areas of the region are grouped into two domains: classified and unclassified, including wildlife reserves (partial wildlife reserve of Nabtrt and Bontioli and the total reserve of Bontioli with a total area of 78,700 hectares) and classified forests (classified forest of Bougouriba, Nabtrt, Koulbi with a total area of 55,000 hectares).

The animal species encountered in the area are: *Erytrocebus patas patas* (monkeys), *Aepyceros melampus* (antelope), *Hystrix cristata* (porcupines), *Lepus capensis* (hare), *Francolinus bicalaratus* (francolin), *Helmeted guineafocol* (wild guinea fowl), *Anas sp* (teal), *Eupodotis melanogaster* (bustard), *Syncerus cafferles brachyceros* (buffalo), *Loxodonta africana* (tltphant), *Phacochoerus africana* (warthog), *Kobus (Adenota) kob kob* (buffalo cob), *Ourebia ourebi quadriscopa* (Ourtbi), *Bucorvus abyssinicus* (great hornbill), (Profile of the Regions in Burkina Faso, 2009/ BLANCA G. et al, 2007).

With all these riches, the region remains favourable to the development of socio-economic activities.

CHAPTER 3: SOCIO-ECONOMIC DATA

The South-West region is composed of several ethnic groups which make it a culturally rich region. It is sparsely populated (38 inhabitants/km^2) compared to the national average of 51.8 inhabitants/km2 (RGPH, 2006). Its natural attributes mean that various activities can be carried out (diversified agriculture and livestock). Its geographical position could easily allow it to maintain trade relations with the neighbouring countries of Cote d'Ivoire and Ghana.

3.1. THE SETTLEMENT OF THE SOUTHWEST REGION

3.1.1. Settlement history

The now legendary resistance of the Lobi to colonisation explains why historians and anthropologists have privileged the study of the colonial fact to the detriment of their precolonial past. The South-West of Burkina Faso was not concerned by the development of political formations that were sometimes reshuffled. It remained for a long time isolated from external influences. Only the Koulango kingdom of Bouna claimed occupation of this area up to the Bougouriba before the XVe siècle ROUVILLE (de) C., (1987).

Concerning the settlement itself, controversies exist between the different authors who have studied the question. They are DELAFOSSE M., (1912) ; LABOURET H., (1932) ; KIETHEGA J. B., (1993), PERE M., (1982, 1992, 2000, 2004). All have given a different chronology of arrival of the peoples, as well as their origins.

Among these authors, we retain the thesis of PERE M., (1982). Faced with the complexity of the historical references, PERE M. specifies: *"in the absence of sufficiently recent and synthetic historical studies on this region, we are forced to resort to sources that are already ancient, among which we willingly refer to the work of Henri Labouret, whose conclusions seem to be the most accurate. Most of the populations that he called "tribes of the Lobi branch" came from present-day Ghana during the 19th century.* In her thesis, she emphasises that: *"this gold-rich region, which is rich in fertile land and has been successively occupied, has been the scene of covetousness and conflicts. Technically more advanced ethnic groups colonised it before the present occupants settled there, judging by the numerous ruins (square, stone enclosures with walls sometimes six to ten metres high). The new settlers found the area uninhabited. They are the Gan, the Koulango, the Tesse or Teguesse, the Pwo, the Djan, the Lobi, the Birifor, the Dagara (lobr and wiile), who came in small family groups, constantly moving in search of game and new land, fleeing famine and slavery.* But PERE M. herself retains the Koulango, the Tessës, the Gan and the Dorossië ɸoroɓë) as the first occupants. The Lobi and Birifor groups would have settledës later, on places that belonged to other ethnic groups.

3.1.2. The distribution of ethnicities

The indigenous population is made up of a multitude of ethnic groups, most of which belong to the great Lobi group. The Lobi themselves occupy the whole of the centre of the region. The Birifor are located in the east, on a strip adjacent to the Gluurean border. To the north, the Dïëbougou-Disssin region is occupied by the Dagara, the Bondigui-Nisseko region by the Djan and the Karankasso region by the Vigg. Finally, the Gan in the Loropëni region and the Dorossïë around Ouo should be mentioned. In addition to these agricultural populations, the immigration of Dioula, essentially traders and weavers, was the last. A few Peuhls transited with their herds during the rainy season. Today there is a strong Moaga community throughout the region (Profil des régions, 2009).

3.1.3. Social organisation

There are two socio-political systems in the region according to ROUVILLE (de) C., (1987):

S centralisë for the Loron, Koulango, Gan and Gbodogo ;

S and not centralised for other peoples. It is a traditional, gërontocratic type of organisation whose structure is marked by clans and lineages. The maintenance of social order is based on respect for the hierarchy of the chief of the village, the concession, the family, the clan, the customary and the village or inter-village.

In the area of land tenure, the chief of the land is dung to settle disputes on the basis of the traditional laws and regulations that govern the societies.

The social organisation is also based on the matriclan with three variants: the maternal parentë, the father's matriclan and the paternal parentë. In the Lobi and Birifor branch, the maternal parentë is predominant. It is a source of frustration nowadays, in the system of inheritance (the nephew takes the inheritance). In contrast, among the Dagara Wiik, inheritance is practised in such a way as to exclude the nephews from taking over the property of the maternal uncle. The only legitimate heirs in this system are the brothers or siblings of the deceased. In the region, women have no right to land (except for the hut fields around the house), which does not facilitate their advancement (Ministere de l'Economie et du Dëveloppement, 2005).

3.1.4. The habitat

In the area of housing, the region recorded in 1998, in terms of the types of materials used for housing construction, more than 96.5% of houses were made of banco (l'Enquete Burkinabe sur les Conditions de Vie des Mënages, 1998). The 2003 survey indicates that 95.2% of houses have earthen walls and 67% of buildings have earthen roofs. The habitat is therefore precarious for a region that is heavily irrigated. Thus, every year, there are homeless people during the rainy season and this often poses problems of hygiene and public health. The hard constructions suggest a certain ease of living (Ministry of Economy and Development, 2005). In urban centres, the number of unserviced spaces is also decreasing. Despite this progress, the 2006 RGPH shows that huts still dominate in rural areas with 54% compared to 18.5% in urban areas. Villas and apartment buildings are almost non-existent in the countryside and poorly represented in the cities.

3.2. THE ADMINISTRATIVE AND DEMOGRAPHIC FRAMEWORK

3.2.1. The administrative framework

According to the (Profile of the regions, 2009), the region has 4 provinces, 4 urban communes, 24 rural communes, 1047 villages and 27 sectors. The province of Poni, with a surface area of 7,514 km^2 (45.45% of the region's surface area), is the largest, followed by that of Ioba (3,272 km^2 , i.e. 19.79% of the regional surface area). The smallest province is Bougouriba with 2,868 km^2 (17.35% of the regional area), followed by Noumbiel (2,879 km^2). Table 4 gives an administrative overview of the region.

Table 4: Administrative organisation of the South West region

Provinces	Number of villages	Number of sectors	Number of rural communities	Number of urban municipalities	Number of cities in 2006	Area in km^2
Bougouriba	133	7|4		1|1	2 868
Ioba	152	7|7		1|1	3 272
Noumbiel	204	5|4		1|1	2 879
Poni	558	8|9		1|1	7 514
Region	1 047	27	24	4|4		1 6533

Source: RGPH 2006.

3.2.2. Demography

The total population of the South-West region increased from 444,323 in 1986 to 485,313 in 1996. This represents an average annual growth rate of 0.74% (RGPH 1996). In 2006, it was estimated at 620 767 inhabitants (RGPH 2006). This represents a growth rate of 2.5% compared to 1996. It represents 4.43% of the country's total population.

Table 5: Evolution of the total population of the region by province in 1985, 1996 and 2006.

	Population in 1985		Population in 1996		Population in 2006		TAAM between 1996 and 2006 in %.
Provinces	Workforces	by report to the region	Workforce	% compared to the region	Workforce	% compared to the region	-
Bougouriba	65 093	14,6	76 498	15,8	101 479	16,35	2,9
Ioba	143 750	32,4	161 484	33,3	192 321	30,98	1,8
Noumbiel	43 726	9,8	51 431	10,6	70 036	11,28	3,1
Poni	191 754	43,2	195 900	40,4	256 931	41,39	2,7
Region	444 323	-	485 313	-	620 767	-	2,5

Source: RGPH, 1996 and 2006 INSD - (Data not required).

> **Population structure by sex and age**

The gender structure of the population reflects the population of the country. Women in the South West region outnumber men. In 2006, according to the RGPH, for a total population of 620,767, there were 321,503 women (51.79%) compared to 299,264 men (48.21%). This numerical superiority of women is noticeable in all the provinces of the region, particularly in Poni (52.38%). As for the age structure, it shows the extreme youth of the region's population. The population under 25 years of age represents 63.73% of the region's population. This young population is an asset for the region if it is employed.

> **Demographic movements**

Natural movements

The South-West region has a lower rate of iKiUilitaty than the national average in 2006 (43.3% against 45.8%). In 1996, it was identical to the national average (48.2%), (RGPH 1996). The fëconditë svntlietic index (ISF) in 2006 shows that women in the region have an average of 6.4 children during their fëcond life. This is down from the 1996 average (7.1), but still more ëкyё than the national average of 6.2. In the South West region, over 123 out of every 1000 children born die before the age of one, while the infant mortality rate is 115.3 per 1000 for the country as a whole. The maternal mortal^ rate is also ëкyё in the region (316.2 per 1000), (RGPH, 2006).

Migratory movements

Migration movements are low in the South West region. The 2006 RGPH data show that despite its proximity to two relatively rich neighbouring countries, individuals remain attached to their place of birth. Less than 15% of natives reside outside their original birthplace (13.5%). There is a disparity between the provinces. The province of Bougouriba has the highest rate of natives outside its territory (21.3%). On the other hand, Loba is the province where the population is almost toothless: only 7.1% of natives are outside the province according to the RGPH, 2006.

3.3. EDUCATION AND HEALTH

3.3.1. Education

The level of education of the population of the South-West is often consideredërë as better compared to the rest of the country. 16% of people aged over 15, can read and ёеnгe in any language compared to 25% at national level according to the RGPH-2006. At provincial level, the rate is 18.6% for Bougouriba, 17.3% for Ioba, 15% for Poni and 11.4% for Noumbiel. As regards primary school enrolment, the Gross Enrolment Rate (GER) is 48.7% and the National Enrolment Rate (NER) is 37.7%.

The secondary school enrolment rate is low at 16%. These low rates may be limiting factors for the adoption of good drinking water, hygiene and sanitation practices in the region.

3.3.2. Health

In the South-West, the health personnel available in public health facilities is below the standards recommended by the World Health Organisation (WHO), especially for doctors. Indeed, the WHO recommends a ratio of one doctor for every 10,000 inhabitants, whereas in 2008, the average for the region was one doctor for every 72,487 inhabitants, compared to a national average of one doctor for every 31,144 inhabitants. The number of nurses is fairly satisfactory; one state nurse for 3,954 inhabitants and one registered nurse for 5,057 inhabitants. The rate of attendance at health facilities is (50%) compared to a national average of (51.8%), RGPH-2006.

3.4. THE ECONOMY OF THE SOUTH-WEST REGION

The economy of the South-West region is mainly based on the primary sector (agriculture, livestock, fishing, wildlife and forestry). The analysis made by the studies of the regional development plan in 2005 shows that the contribution of the primary sector to the gross domestic product of the region is the most important; it is 57.2%. That of the secondary sector is 26.8% against 16.8% for the tertiary sector (Ministry of the Economy and Development, 2005).

3.4.1. Agriculture

Agriculture has enormous assets and occupies about 85% of the population. Indeed, the region still has fertile land compared to the rest of the country. In addition, the rainfall is acceptable. However, agriculture is still not very open to the market economy and its means of work are still rudimentary on the whole. The weakness of technical supervision, illiteracy and the difficulty of accessing agricultural credit partly explain this situation. However, the region was self-sufficient in food in 2012. The rate of coverage of needs according to the cereal balance sheet during the 2012-2013 agricultural season places two provinces of the region (Bougouriba and Poni) in a balanced situation with a coverage rate of between 90 and 120% and the other two provinces (Noumbiel and Ioba) in a surplus situation with a coverage rate of over 120% (the fortnightly newspaper, L'Evenement, No. 250 of 25 February 2013, p10).

> **Food crops**

Crop production consists mainly of maize, small millet, white sorghum, red sorghum and rice. Their production has increased over the last three years, especially for maize and rice, which have averaged 71 000 and 11 000 tonnes respectively. These figures are shown in Table 6 below (2000 to 2011).

Table 6: Cereal production from 2000 to 2012

YEAR	Mil	But	Rice	White sorghum	Red sorghum
2000-2001	50 745	42 316	7 510	-	-
2001-2002	41 906	36 996	4 621	-	-
2002-2003	44 575	40 812	4 529	-	-
2003-2004	44 980	43 977	4 823	36 901	41 223
2004-2005	55 956	45 538	2 955	45 627	36 807
2005-2006	43 729	39 210	6 470	28 166	36 557
2006-2007	49 061	53 151	11 662	32 954	40 045
2007-2008	49 065	33 308	8 802	42 540	38 993
2008-2009	70 238	68 990	7 803	56 493	52 025
2009-2010	71 103	70 354	11 934	54 324	50 942
2010-2011	70 007	90 829	10 632	68 815	60 729
2011-2012	46 515	71 439	13 504	42 011	46 592

Source: Agricultural statistics, 2012. - Lack of data

Other food crops include yams, which account for more than 52.2% of national production and make the region the leading producer of this crop. Souchet, potato, nitbt and voandzou are also produced in the region.

Cash crops

Cash crops are mainly represented by cotton, groundnuts, sesame and soybeans. Cotton and groundnut production increased from the 2008-2009 to the 2011-2012 crop year, while sesame production fluctuated from year to year. The same is true for soybeans, which reached a production of over 1,000 tonnes in the 2009-2010 and 2010-2011 crop years.

Table 7: Cash crops from 2000 to 2012

YEAR	Cotton	Sesame	Peanut	Soya
2000-2001	20 573	7	11 070	172
2001-2002	20 659	439	8 969	419
2002-2003	12 430		9 878	169
2003-2004	16 063	24	9 407	311
2004-2005	14 075	42	10 256	416
2005-2006	15 053	27	7 611	467
2006-2007	18 488	28	9 528	562
2007-2008	10 928	73	6 567	269
2008-2009	51 488	186	20 864	866
2009-2010	43 136	93	23 876	1 239
2010-2011	60 524	283	28 306	1 005
2011-2012	32 986	97	23 856	508

Data source: agricultural statistics, 2012

> **Market gardening**

The main vegetable crops are: onions, lettuce, tomatoes, chillies, cabbage and aubergines. Vegetable production is mainly organised in the lowlands and around some dams. It is not very developed in the region.

3.4.2. Breeding

It is the second most important sector in the economy of the South West after agriculture. The indigenous farmers are both farmers and herders. It is extensive and traditional with local species of low productivity. However, livestock farming is a promising activity given the potential it offers. These are: the large availability of grazing land, the existence of fairly well-structured groups, and the diversity of sectors. The South-West region has very large commercial and transhumance movements of livestock. The table below shows the number of livestock and poultry in the South West region from 2003 to 2012.

Table 8: Livestock production from 2003 to 2012.

YEAR	Cattle	Sheep	Goats	Equines	Asins	Pigs	Guinea fowl	Chickens
2003	275974	193475	447560	424	2131	251176	426801	1179007
2004	281492	199278	460985	427	2172	256197	935959	3135446
2005	287120	205254	474812	430	2215	261319	1250804	452789
2006	292860	211410	489055	433	2258	266543	466371	1288327
2007	298714	217750	503724	436	2302	271872	480360	-
2008	304686	224281	518834	439	2347	277308	494769	1366783
2009	310778	231007	534396	442	2393	282851	509610	1407785
2010	316992	237935	550426	445	2439	288506	524896	1450017
2011	323330	88009	206649	174	2487	294274	540641	1493515
2012	329795	252423	583944	451	2535	300157	556858	1538319

Source: Direction Generale de la Prevision et des Statistiques de 1 Elevage (DGPSE, 2003-2012).
-Lack of data

There is a slight increase in the number of heads of the different species of livestock in the region. The animals slaughtered in the region are intended for local consumption. The hides and skins are the raw material for handicrafts. People use these by-products to make bags, quivers, tom-toms and wooden chairs (ties). It is estimated that only one third of the production is exported. Exports of hides and skins are similar to those of live animals, i.e. trade takes place outside formal organised structures, making it difficult to control the real flow of trade in livestock products.

3.4.3. Fishing and hunting

Fishing in the South-West region is practised all along the Mouhoun River, but also in the sub-catchment areas of the Bougouriba and Poni rivers, which contain large bodies of water, as well as at the Poniro, Bapla and Batie dams. The main fish species encountered are : *Clacias anguillaris* (catfish), *Hydrocyon brevis* (dogfish), *Distichodus rostratus* (false captain), *Malopterus electricus* (electric fish), *Heterotis niloticus.*

Hunting activity in the South West is not well developed. About 16 licences are issued each year for sport hunting. Traditional hunting is also practised, with licences issued to members of the village association (Ministry of Economy and Development, 2005).

3.4.4. Industry, mining and crafts

The industry

Industrial activity is almost non-existent in the region. The region has only one SOFITEX cotton ginning plant, located in the town of Diëbougou.

The mines

In terms of mining, prospecting has shown the existence of certain minerals in the region. Gold panning is intensively practised in certain areas, notably a Gueguere in Loba, Bondigui in Bougouriba, as well as in the alluvial deposits in the departments of Gbomblora, Kampti and Nako in Poni and in the province of Noumbiel. According to a prospective study carried out in February 2001 by the Direction Générale des Mines, there are copper deposits at Dienemera and Gongondy totalling 24 million tonnes of ore a 0.8% Cu.

The craft

Handicraft is a secondary activity carried out by the population. It includes art, production and service crafts. Art crafts express the past or present culture and are mainly manifested through pottery and sculpture. In the region, pottery and basketry are purely female activities. Blacksmithing, woodworking, balafon making and granary construction are male activities. Service or production crafts encompass all activities providing maintenance, carpentry, etc. This informal sector lacks professional qualification and is not well structured (Ministère de l'Economies et de Developpement, 2005).

3.4.5. Tourism and the hotel industry.

The South West is full of tourist sites, the most important of which is the Poni Museum, which is both an ethnographic and traditional habitat museum and a living museum or ecomuseum.

Table 9: Tourist sites in the South West.

Province	Department	Location of the site	Name of the site
Noumbiel	Batie	Municipality of Batie	French military cemetery
			The tomb of Da Mar (founder of Batie), the remains of the first secondary school in Burkina, the house inhabited by President Felix H. Boigni.
loba	Dano	Dano	The Loba hills; Dagara habitat.
	Zambo	Djikologo	The Djikologo cave.
Bougouriba	Diebougou	Diebougou	The cave of the French army.
	Dolo	Dolo	The Djan traditional healers.
Poni	Gaoua	Commune of Gaoua	The provincial museum; the cave of Tambili hill.
	Loropeni	Loropeni Obire	The ruins of Loropeni; the sanctuaries of the Gan kings.
	Gbomblora	Gbomblora	The big house of Da Bindoute.
	Kampti	Kampti	The hills to be sliced.

Source: Directorate General of Economy and Development (DRED) /SO/2005

In terms of hotels, all provinces in the region have relatively low-capacity accommodation units. The South West region has two hotels, thirteen hostels, eight camps and nine accommodation and reception centres (Profil des régions au Burkina Faso, 2009).

PARTIAL CONCLUSION

The environment of the South-West region has quite favourable conditions for wealth production, especially in the agricultural and livestock sectors. However, the abuse of this potential could jeopardise the future of the gënërations if nothing is done. There is a continuous degradation of natural resources (deforestation, silting of rivers, soil degradation and biodiversity). This situation is linked to the extensive production methods adopted in the region for both livestock and agriculture. The school enrolment rate is low, as is the attendance rate at health centres. The standard of living and quality of life is low. Drinking water, hygiene and sanitation practices are strongly influenced by these factors.

PART TWO

PRESENTATION OF THE RESULTS

40

CHAPTER 4: DRINKING WATER AVAILABILITY IN THE SOUTH-WEST REGION

The amount of rain that falls in the southwest is among the highest in Burkina Faso (900 to 1300 mm per year). However, the presence of both natural and artificial water points is limited. This could be explained by the Birimian structure that dominates this part of the country and the insufficient construction of hydraulic infrastructure.

The natural and artificial water points are: rivers, shallows, streams, dams, wells, boreholes and AEPS. The national inventory of water points carried out in 2005 updated the 1996 inventory and established an exhaustive inventory of the real situation of drinking water supply and sanitation. Our study, although not exhaustive, allowed us to re-do the inventory of these water points in our sample villages (see list in tables 2 and 3) and to geo-reference them with the help of the GIS tool, which allowed a spatial visualisation of the different hydraulic structures.

4.1. THE DIFFERENT ACTORS AND THEIR ROLES

4.1.1. Town halls

The 2009 decree on the transfer of water and sanitation responsibilities to the communes replaces the State with town halls, which now have the task of providing all the villages in their administrative district with drinking water, hygiene and sanitation facilities. In the twelve town halls visited in the course of this study, we were able to talk to the various officials. All the town halls acknowledge that there is a lack of drinking water points.

However, little action has been taken by the latter to provide the villages with infrastructure. Indeed, none of the town halls, except for those of Batie, Dano, Diebougou and Koper, has been able to provide any concrete or future infrastructure. The main reason given is the lack of means. They have low revenues in terms of tax collection. The sectoral WATSAN Communal Development Plans (CDPs), for those town halls that have them because they do not all have them, are oriented towards development partners and the State. In fact, they are designed to request funding from partners. According to the DGAEUE, only six of the twenty-eight communes in the region had a WATSAN sectoral CDP before 2010. The actions of these municipalities are limited to the repair and rehabilitation of existing boreholes. Indeed, the construction of a new borehole is expensive (between 2,600,000 and 5,500,000 F cfa, according to the World Bank report, 2007) compared to the financial resources of the town halls. The cost depends on the lithological nature and depth of the water table. The construction of new boreholes is the work of NGOs and associations that support these communes in this field. The focal points in charge of making an inventory of modern water points are agents of the Ministry of Agriculture and Food Security. It is therefore difficult, under the current conditions, to significantly increase the level of drinking water infrastructure in the South West region, which remains low by any standards.

4.1.2. Development partners

NGOs and associations have always accompanied the state in the construction of drinking water infrastructures. Today, they directly accompany the communes. The NGOs and associations present in the South West region and which intervene in the field of WATSAN are

- Valorisation des Ressources en Eau dans l'Ouest du Burkina (VREO) is an NGO funded by the European Union; it carried out an AEPS in Bondigui;

- VARENA-ASSO: in collaboration with WaterAid, helped the commune of Dano to develop its sectoral CDP for drinking water and sanitation;

- Programme National de Gestion du Terroir Phase Two (PNGT2): this programme realised two boreholes in the rural commune of Midebdo; the rehabilitation of two boreholes in Dadoune in the rural commune of Niego; a borehole in Dianle in the rural commune of Ouessa; a large-diameter well in Complan, two large-diameter wells in Balembar and two modern small-diameter wells in Yo in the commune of Dano;

- PLAN Burkina: wants to set up an AEPS in Midebdo and is the partner of the commune of Batie in AEPA;

- SOS SAHEL: this NGO has built a large diameter well in Balembar, a village in the commune of Dano in the Ioba;

- WaterAid: In addition to its participation in the elaboration of the communal sectoral development plan, AEPA has rehabilitated a borehole in Tambiri, a village in the commune of Dano.

There were five NGOs and one local association in the area.

Table 10: The different NGOs per commune

Provinces	Municipalities	NGO
Bougouriba	Bondigui	VREO
	Diebougou	-
Ioba	Dano	VARENA-ASSO, WaterAid, PNGT, SOS SAHEL,
	Gueguere	PNGT, VARENA-ASSO
	Koper	-
	Niego	PNGT
	Ouessa	PNGT
	Zambo	-
Noumbiel	Batie	Plan Burkina
	Midebdo	Plan Burkina
	Legmoin	-
Poni	Boussera	-

Source: field data 2012-2013

The NGOs, broken down by province, show that 100% of partners are active in Ioba, 20% in

Bougouriba, 40% in Noumbiel and no partners in Poni. The percentage for the latter province applies only to our study area, the rural commune of Boussera. The commune of Dano stands out from the others with 80% of partners involved. The preponderance of partners in this commune could be explained on the one hand by the dynamism of the commune's leaders and on the other hand by the WATSAN objectives which are well elaborated and thus easily mobilise the partners. As proof, we note the elaboration of the sectoral WATSAN PCD which is the work of WaterAid and VARENA-ASSO. The dynamism of this commune should serve as an example for the other communes of the region.

4.1.3. Water point management structures

Through the Village Water Supply Programme, the state has sensitised the users to take ownership of the facilities and ensure their maintenance through the establishment of water point management committees and the training of rural mechanics to repair the pumps. The committee relied on funds generated by an initial contribution from the users, which was later reinforced by their contributions or the proceeds from the sale of water. For SAVADOGO A. N., (2012): 'after the first few years of the project, water coverage levels that sometimes reached 80-90% at the end of the project quickly dropped to 60% or even 40%, often necessitating new projects to rehabilitate defective boreholes or pumps'.

As in the rest of the country, boreholes in the South-West suffer enormously from maintenance problems. Maintenance is carried out by rural pump mechanics, formerly trained by the Office Regional de Développement (ORD) and currently by the Ministry of Water, Hydraulic Amenities and Sanitation. In fact, the inorganisation of the communities and the inadequacy of the awareness campaigns among them mean that the populations have not taken ownership of the works. Of the forty villages covered by our study, we were not able to reach enough water point managers (only three water point managers were reached). This is due to their absence. They do not sell water as their main activity, so they carry out other activities to meet the needs of their families. In other cases, the managers are non-existent. Table 11 shows this situation.

Table 11: Managers of water points encountered in the South West

Provinces	Municipalities	Villages	Number of water point managers
Bougouriba	Bondigui	Bondigui	1
Ioba	Ouessa	Ouessa	1
Poni	Boussera	Kpara	1
Noumbiel	-	-	0

Source: Field data 2012-2013 - Lack of data

According to the table above, in Noumbiel, no water point manager could be reached. In the villages where there is an organisation, the managers are not present and the boreholes are not padlocked against those who might use them without paying. The inhabitants have access to water for free.

Another problem is the poor quality of the equipment used for construction or repair. In addition, spare parts are not available close to the population. In case of breakdown, they go back to the sources of non-drinking water while waiting for a goodwill person or the State to repair the well. In Bile, in the rural commune of Gueguere, there is zero coverage of drinking water because the 4 boreholes in the village have broken down.

Photo 1: Pump breakdown in Kalambouro

Author : MALO Y., 13/12/2012

This borehole is the only one in Kalambouro and it has not been working properly for months. To repair it, the material and the mechanic have to come from Batie or Gaoua. The first user of the day has to bring water from home in order to bring the water up. In other words, the water will not come out of the pump. This is also necessary when the two uses of the pump are spaced. As shown in photograph 1, the pump must be dismantled and water put in before it can be operated. This practice calls into question the potability of the water from this borehole. Indeed, the water brought from the house is not free of contamination and will in turn contaminate the borehole water.

4.2. TYPOLOGY OF DRINKING WATER POINTS IN THE SOUTH-WEST REGION

Two sources of water are exploited in the southwest: groundwater and surface water. Surface water sources are limited to rivers and dams. Groundwater is exploited through boreholes, traditional wells, modern small and large diameter wells and AEPS. In the 40 villages, we counted a total of seven hundred and seven (707) water points of all types.

We counted 632 functional and 75 non-functional water points. 64% of the non-functional water points are sources of drinking water including boreholes and standpipes. These different water points are shown on the map below. The majority of these water points are located on the armourstone glacis and on the armourstone level.

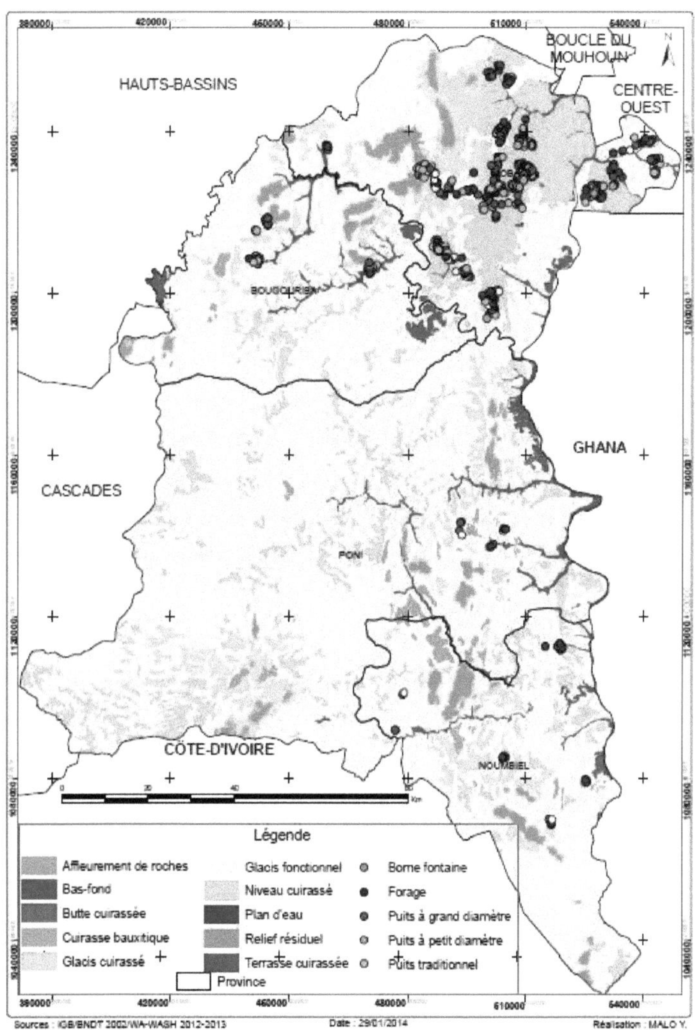

Map 5: Distribution of drinking water points

The observation of map 5 shows the unequal distribution of water points with a high density in the Ioba province and a low concentration in the Noumbiel and Poni, while the Birimien model covers the whole region.

4.2.1. Drilling

The inventory of boreholes in the southwest shows that they represent 26% of the existing water points in the region. 78% of these boreholes are functional and 22% non-functional. School and health boreholes represent 9% of functional boreholes. The level of drinking water infrastructure is therefore low. In addition, breakdowns hinder the proper functioning of these infrastructures.

Today, according to SAVADOGO N. A., (2012), the notion of village hydraulics has ëvoluë and we speak more of district hydraulics in state policies. It is a policy that goes down to an even smaller scale than the village. But the policy of village water supply, which seems satisfactory overall in view of the 50,000 water points available, hides disparities. It is even tempting to say that all villages have at least one modern water point. In the South-West, especially in Noumbiel, an insufficiency, or even an absence of drinking water points in some villages is noticed. According to the Mayor of the rural commune of Midebdo, out of the 52 villages in his commune, 18 villages do not have a drinking water point. Map 6 below shows their distribution in the region.

The largest number of boreholes is found in Ioba. Most of the communes in this province have a fairly good coverage of boreholes. On the other hand, the lowest number of boreholes is found in Noumbiel and Poni, where all communes have a high level of need (Map 6 below). This is consistent with the logic that the further away from the centre the less infrastructure there is. The authorities in the different localities do not make the same efforts to provide their localities with drinking water infrastructure.

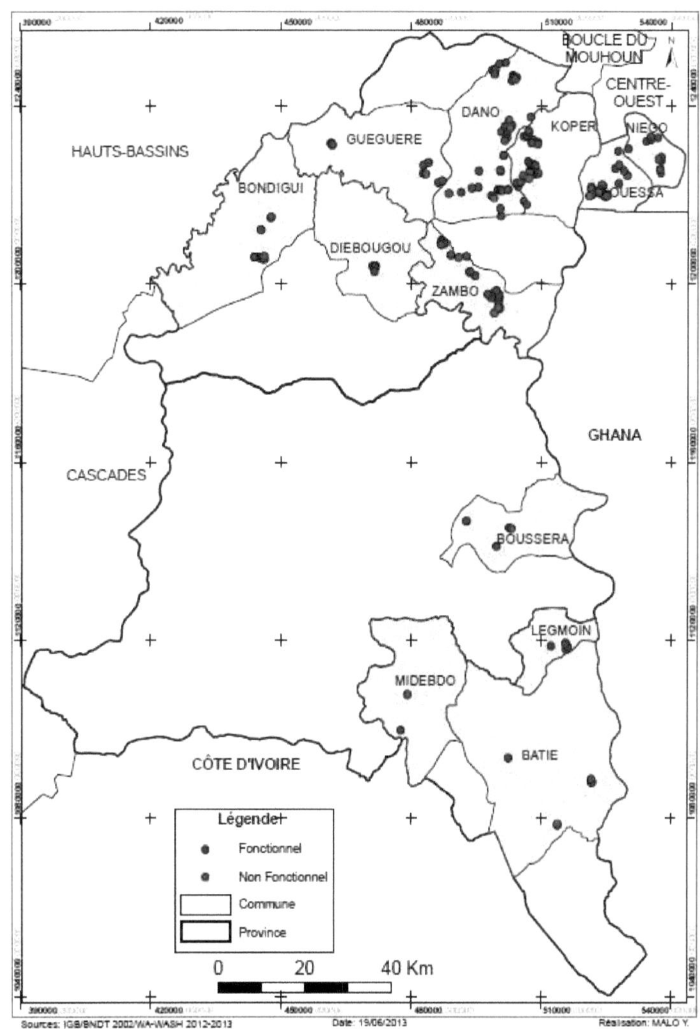

Map 6: Boreholes in the region

4.2.2. Standpipes

These are simplified drinking water supply infrastructures. They reinforce the availability of already existing drinking water infrastructure. They are located in the main towns of urban and rural communes. In 2005, at the national level, there were 472 simplified drinking water supply networks (AEPS) and autonomous water stations (PEA) with 311 functional and 161 non-functional, i.e. 34% (Ministère de l'Agriculture, de l'Hydraulique et des Ressources Halieutiques, 2006).

In the case of the South West region, there is a notable under-equipment of this type of infrastructure. Of the twelve rural communes, four chief towns have AEPS, i.e. 33% of rural communes. 25% of these communes have non-functional AEPS. According to our analysis, standpipes represent 2% of the region's water points. Half of these are not functional. They are found in Bondigui, Koper, Bapla and Ouessa. The Ouessa AEPS is not functional due to management problems and has been converted to a human powered pump (HPP). The use of AEPS requires rigorous management and a contribution from the population. It is a costly device that most often uses a generator or solar panels to raise water. AEPS are not well managed in the region, as evidenced by their level of functionality. Map 7 gives an overview of the standpipes in the region. It shows a virtual absence of standpipes in the communes of Noumbiel and Poni, and a significant number in the communes of Ioba province and Bougouriba province. All the standpipes in the rural commune of Koper are functional, while in Ouessa and Bapla, none are.

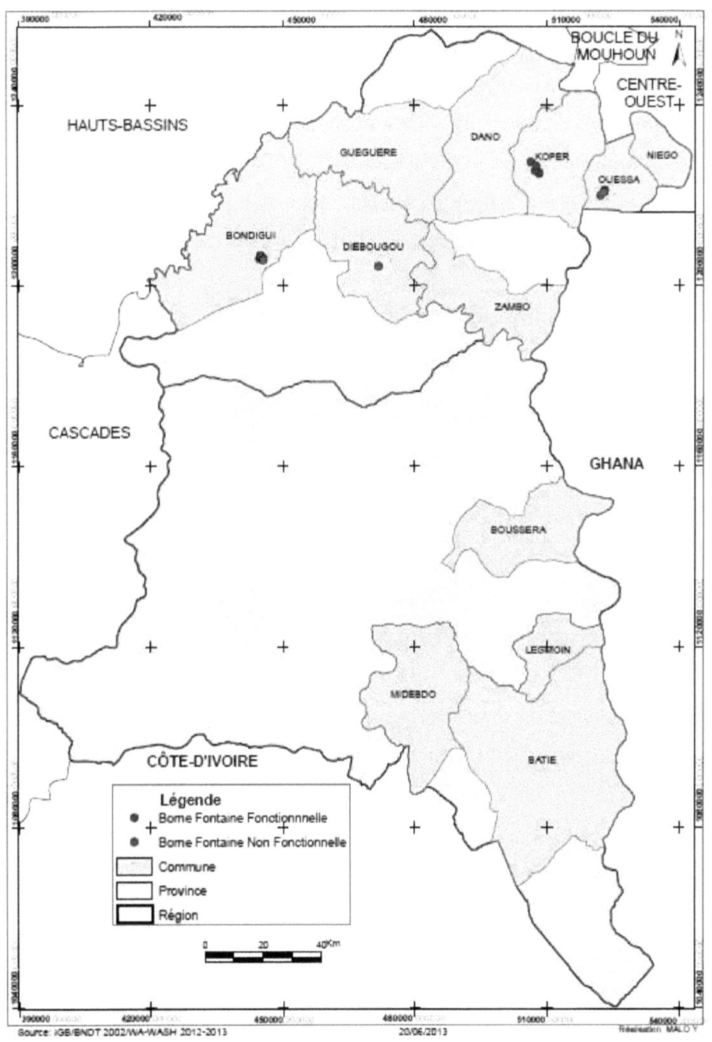

Map 7: Distribution of standpipes in the South West

4.2.3. The wells

Modern wells

The first water seminar held in Burkina Faso in 1976 decided on the priority choice of boreholes over wells in rural areas because of their temporary nature and the risks of contamination which are high (SAVADOGO A. N., 2012). Modern wells account for 30% of the region's water points, of which 8% are small diameter wells and 22% are large diameter wells. These wells are often temporary and only temporarily solve water problems (during the rainy season).

The average depth of modern large diameter wells is 22m. The shallowest (7.5m) is found at Bapla in the commune of Diëbougou. One of the few large water reservoirs (dam) in the region is located here. This dam could explain this rise in the water table. The deepest dam is located in Zambo and is 62 m deep. This depth can be explained by the relief which is uneven. The summits and plateaus are the areas of steep slopes, therefore with dominant surface flow. The infiltration is very fine and the marks of intensive erosion are generally observed.

The average depth of modern small-diameter wells is 17m. The shallowest (7 m) is found in Dalgane in the rural commune of Koper and the deepest (35 m) in Complan in the commune of Dano. This type of well is only found in the province of Ioba. The temporary nature of these wells is explained by their lithological nature. The thinner surface layers retain less water which is mostly drained to neighbouring countries. Map 8 shows the distribution of modern wells.

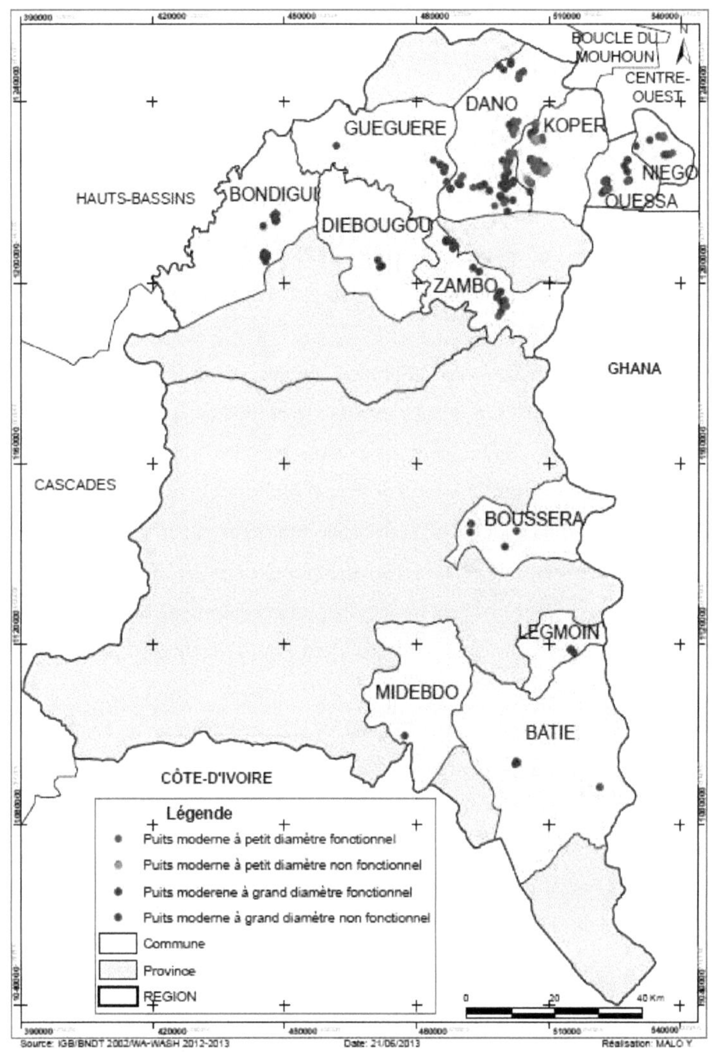

Map 8: Distribution of modern wells in the South West region

Traditional wells

They are among the most used water points in the region. They represent 34% of the water points. The absence of a well cap means that rainwater drains into them. These wells are shown on Map 9 below. Most of the traditional wells are functional. In the villages of Noumbiel and Poni, they are absent. The province of Ioba, despite its good coverage of drinking water, has the highest number of traditional wells.

These traditional wells, as well as boreholes, AEPS and other types of wells, are highly dependent on groundwater that needs to be recharged. They are recharged in three different ways:

- direct recharge by homogeneous infiltration. Rainwater infiltrates directly into the soil and slowly progresses into the subsoil as a moisture front (diffuse recharge).

- direct recharge by preferential routes: heat waves, fractured zones, quartz veins (recharge by preferential routes).

- direct supply: this is done by runoff water concentrated in and around local topographical depressions (lowlands or marshes) and regional depressions (alluvial valleys). This recharge of the water table can be done by a wet front or by preferential ways (MARCELIN J., 1971).

In the region, recharge is difficult because the steep slopes accelerate the flow of rainwater and the low soil thickness does not allow the water table to hold a significant amount of water. This would explain the depth of some of the wells in the region as well as the temporary nature of others according to our analyses.

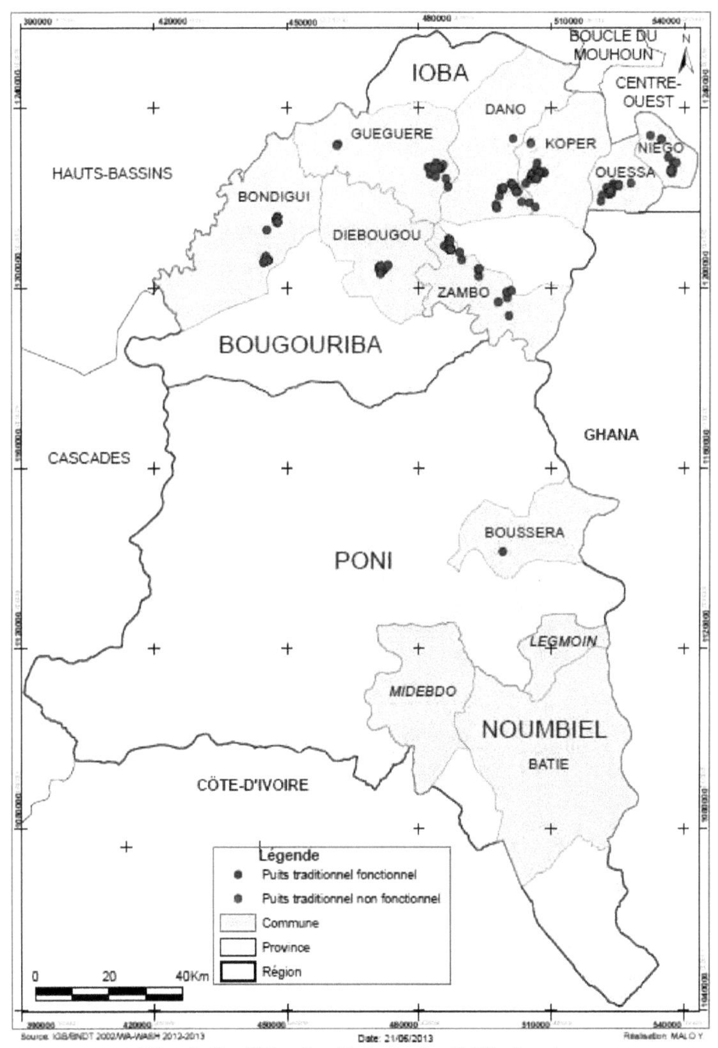

Map 9: Traditional wells in the South West region

4.2.4. Surface water

It is the water of dams, rivers, shallows and marshes. This water is very sensitive to pollution. Contamination is mainly linked to the presence of houses near water points. In addition, the multiplication of works, the use of pesticides in certain irrigated areas and discharges of all kinds, in particular heavy metals and toxic substances, can multiply the risks of degradation of the quality of the water, locally, but also by transport in the water courses. In the south-west, it is a source that is frequented in the same way as the other sources. Indeed, surface water represents 5% of water points and in some villages it is the main source of water supply for the population. With the cultivation of cotton using pesticides and chemical fertilisers, this water is reportedly polluted in the region. Another source of pollution is traditional gold panning which uses chemicals (cyanide, mercury) in the margins of environmental standards. In photograph 2 a, the granite outcrops on the bed of a watercourse and prevents infiltration and the dead leaves that fall on it pollute it. In photograph 2b, the river is less than 200m from cotton fields which pollute it by seepage or pesticide run-off. In both photographs, the method of sampling is the same (entering the water and collecting it).

Photo 2 a: Examples of surface water abstraction sources

Author : MALO Y. (13/12/2012, Koubeo-Djoulo)

Photo 3 b: Example of surface water abstraction sources

Author: MALO Y. (05/01/2013 in Djikologo).

4.2.5. Drinking water coverage

The drinking water coverage rate is an indicator of the degree of adequacy between the water needs of the population and the available resources. These resources should in principle be permanent and meet potability standards with good accessibility.

Drinking water coverage rates were calculated according to national standards that stipulate one borehole/standpipe for up to 300 inhabitants and at least one borehole for every administrative village regardless of population size and accessibility within a one-kilometre radius (PCD, WATSAN Dano, 2011). Boreholes belonging to education, health and other non-community structures are not included in the calculation of village rates. The results at provincial and regional level are presented in the map below. There is an unevenness in the rate of access to drinking water between provinces on the one hand and between communes on the other.

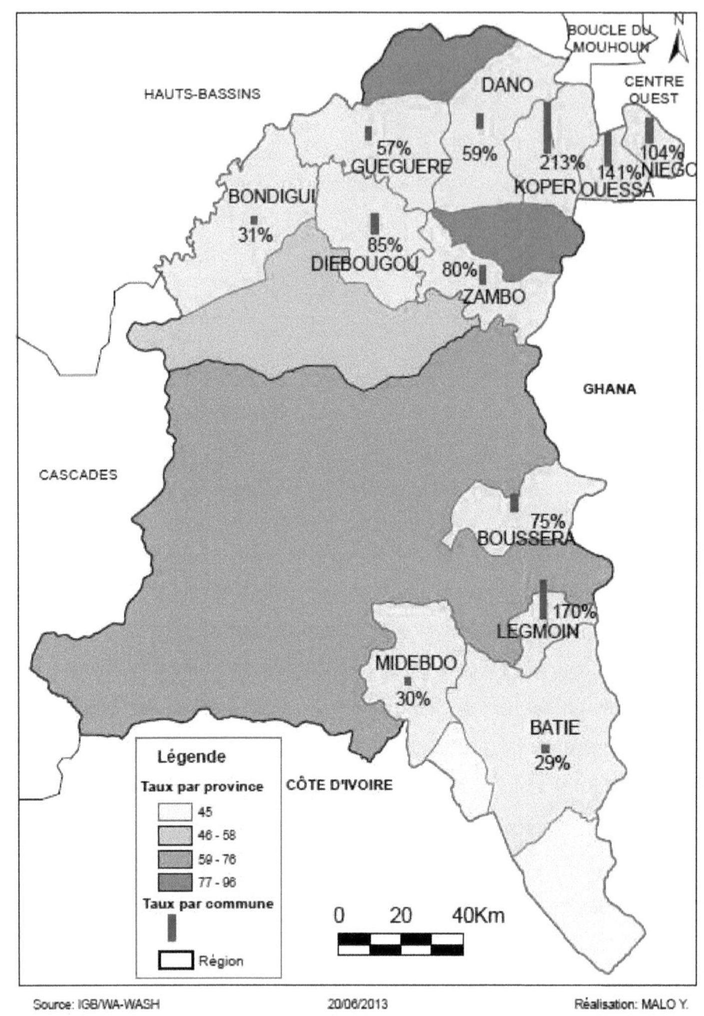

Map 10: Drinking water coverage by province and commune

The results on the map show that only Bougouriba and Noumbiel, with 58% and 45% coverage respectively, do not have an access rate close to the MDG (76% coverage). Loba has a higher rate than the MDGs (96%) and Poni has a coverage rate of 76%. Disparities become significant at the communal and village levels.

At the commune level, we find that those of Bondigui, Batië and Midebdo have less than 50% coverage. On the other hand, the commune of Koper has a coverage of over 300%. At the village level, some localities have almost no coverage (Koubeo- Djoulo, Kpanhila, Bile, Torkouora). Graph 2 shows this distribution.

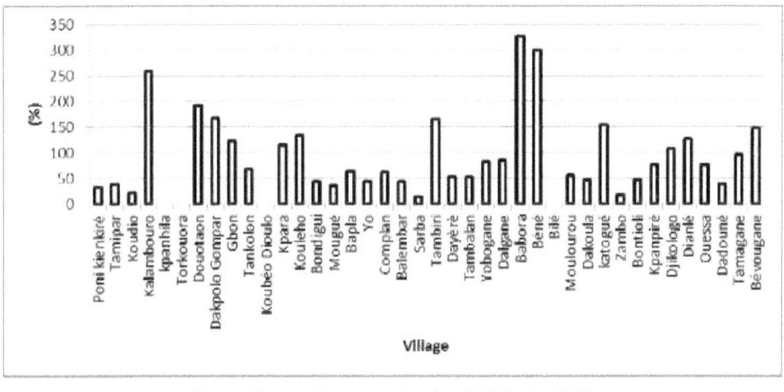

Source : Données de terrain (décembre 2012-janvier 2013)

Figure 2: Drinking water coverage by village.

CHAPTER 5: ACCESS TO DRINKING WATER

In this study, we define accessibility to drinking water as the availability of water in a short time, at a short distance and at low cost from a pump or a permanent standpipe. In the Birimian environment, accessibility becomes more complex due to the peaks to be climbed and the difficulty in using certain means of transport such as rickshaws which allow the use of several containers in one lane, thus increasing the volume of water, and bicycles which facilitate the journey.

In Burkina Faso, access to safe drinking water remains an ongoing struggle and the authorities have yet to find definitive solutions. In 2005, 60% of the rural population had "reasonable" access to drinking water, as defined by WHO/UNICEF and taken into account by the standards, criteria and indicators in our country (PN- AEPA, 2006). It also mentions an inequality between localities.

5.1. THE USE OF DRINKING WATER SOURCES

Following our surveys, it appears that only 36% of households use boreholes and 2% use standpipes (found in some rural commune capitals). In addition, 12% of households combine the use of boreholes with that of wells (traditional and modern) and rivers. Boreholes and standpipes are increasingly the only drinking water sources in rural areas. Indeed, in the field, we have noted an absence of newly constructed modern wells. The modern wells are less so because they are not covered and therefore dirty, and the State itself no longer encourages their construction. N., (2012).

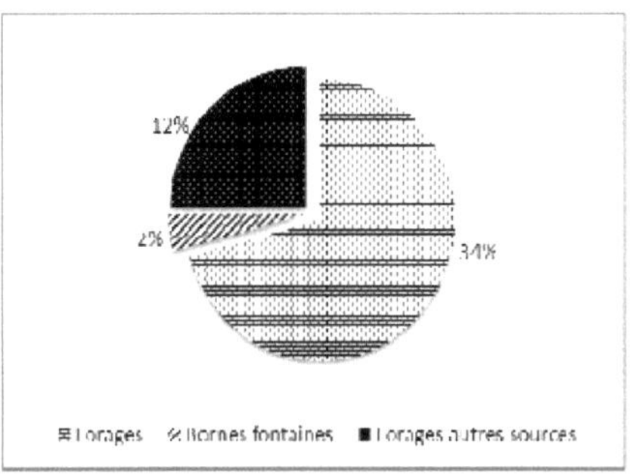

Source: Field data (December 2012-January 2013)

Figure 3: Populations using drinking water sources

These data show that less than 50% of the population use drinking water sources.

5.2. ACCESS FACTORS TO DRINKING WATER IN THE SOUTH-WEST

> **Distance in access to drinking water**

According to the Direction Générale du Sud, quoted by GALBANI S. R. P., (2011), access to drinking water means: "the ability to have running water at a distance of no more than 100 m". For WHO (2003), in terms of distance, access to drinking water means the existence of a permanent drinking water source within 200m of the household. The value adopted by the MDGs and retained by the INSD for the analysis of household living conditions is 1000 m or 30 minutes. Generally speaking, housing is scattered in rural areas in the South West. It is much more scattered in Noumbiel and Poni and less so in Bougouriba and Ioba. Of the 38% of households that actually use drinking water sources, 80% travel less than 500m, 17% one kilometre and 3% more than one kilometre. These analyses show that the greater the distance, the less the population travels, and the less the distance, the more the population uses the water source. With scattered settlements, distances mean that people do not visit boreholes. These results show that distance could be a determining factor in the use of drinking water sources, especially as 38% of the population actually use them. Our surveys also show that 34% of households do not have a source of drinking water close to them. The impact of distance is most apparent in the sketch of the village of Kpara in the rural commune of Boussera.

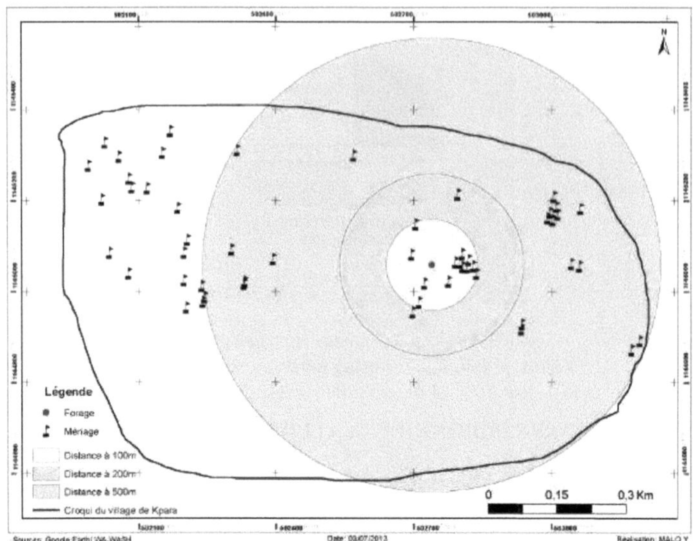

Map 11: Sketch of the distance of households to the drinking water source in Kpara.

There is a large proportion of households that do not have access to safe drinking water according to WHO (2003) and Southern Command standards. Not all households located more than 500 m from the water point go there, as 80% of them travel less than 500 m. The analysis of this map shows that 64% of households can access water because they are located 500 m from the borehole. On the other hand, 36% of the households are not likely to have access to drinking water because they are located more than 500 m from the borehole.

> The Price of Access to Drinking Water

In rural areas, the price of water is critical to ensure the maintenance and sustainability of waterworks. The elasticity of water demand is negligible, if not zero. This means that prices have little influence on demand and the quantities of domestic water consumed per inhabitant vary little, whether prices increase or decrease (PROST A., 1996). In the south-west, of the households that obtain water from boreholes, 49% have free access to drinking water and 51% pay. For the latter, the collection of their fees is not as effective. In almost all the pumps visited during our data collection, no water point manager was present. There is therefore a laxity in the collection of fees at the pumps. Despite this, some people are not prepared to pay their dues. For example, in Kpara, the water point manager tells us that not everyone has access to water from the borehole. Those who refuse to pay their dues are excluded. Photograph 3 shows a woman whose household is excluded from the borehole because it does not pay its contribution. It should be noted that this price varies from one borehole to another. For some boreholes it is 200 f per month, for others it is 500 to 1000 f per year. The price at the source is less, but it is 10 f per 20 l can.

Author: MALO Y. (14/12/2012).

His household is about 800 m from the borehole. This factor may also explain the refusal to pay the contribution, as the household would not be willing to travel such a distance to get water. Another reason is given by the water point manager. The manager says that the neighbourhood where the borehole is located is a source of open conflict with the other neighbourhood. It is December and the river has already dried up. The silting up of the river bed has accelerated this drying up. According to this woman, she is digging the bed as the water table falls. Water erosion is said to be active and intensive in the locality. This is done at the expense of the summits.

Another, slightly different, situation is observed in Kpanhila in the rural commune of Midebdo, where the only drinking water point in the village has been out of order since 2000. When the only borehole breaks down, people have no choice but to drink water from the rivers or go to a neighbouring village to get drinking water. This is a village where there is not a single well (modern or traditional). The lack of wells could be explained by the depth of the water table or by the nature of the rocks. The price is in this case a determining factor in the access to drinking water. It is a deterrent for the less affluent or the less aware on the one hand, and on the other hand, it ensures the sustainability of the work if the contributions are well managed.

> The time factor in access to drinking water

In the South West region, people were not able to estimate the time needed per day to collect water. However, using the observation grid associated with the household questionnaire, we found that households spend at least 3 hours per day collecting water. If people were to get water only from drinking water sources, this time would have to be doubled, as a distance of

61

more than two kilometres can separate two districts in a village. Balembar is a village that best illustrates this situation. It has nineteen neighbourhoods that are between one and eight kilometres from the centre. The village has only 14 boreholes, two of which are out of order. Of the 12 functional boreholes, five are in the centre and the other seven are spread over the remaining 18 wards. The same situation is observed in other villages, notably Dakoula in the rural commune of Gueguere, which has 10 districts with two boreholes. It is even more marked in other villages where one has to go to the neighbouring village to get drinking water (this is the case in Koubeo-Djoulo) which has no borehole and the only large diameter well is 40m deep and temporary. Between two neighbourhoods there may be a hillock, a hill or a river that separates them.

The distances, combined with the volume of the transport container (basin), lead women to make several trips to draw the amount of water needed to satisfy their needs. The WHO standard (adopted by Burkina Faso in the PN-AEPA, 2006) is 20 litres of water per day per person in rural areas, whereas according to our analysis, the average household population is 11 in the south-west. To meet the water needs of a household, 220 litres of water are needed per day. To reach this volume and meet this standard, women are forced to suspend certain activities. Despite this, they are unable to meet the amount of water needed for their households.

If for KOMBASSERE A., (2007), the notion of access to drinking water is an indicator that represents the quantity and quality of water available to a person per day, we can say, in the light of our analyses, that the population of the South-West does not have access to drinking water in its entirety.

5.3. THE USE OF NON-DRINKING WATER POINTS

> Well frequentation

Traditional wells have an important place in society. They are still very much used by the population. Their preponderance is explained by their low cost. Anyone can dig a well in front of their door or in their yard. As for modern wells, their establishment was strongly defended by some voices during the first water seminar in Burkina Faso in 1976, SAVADOGO A. N., (2012). The results of our surveys show that 41% of households use wells. Of these, 36% use traditional wells and 58% use modern wells. For those who use traditional wells, they are exposed to serious health risks. The construction of these wells in some villages in the region is difficult because the geological substratum does not allow it. In Noumbiel and Poni, the lithological nature of the land and the relief do not facilitate the construction of traditional or even modern wells. This is the case in Koubeo-Djoulo where the granite outcrops at river level, while the households are perched on the mountain slopes.

Surface water frequentation

Surface water is used extensively in the South West region. Following our analysis, we find that about 7% of the households surveyed obtain their water mainly from rivers. These are the most polluted sources of water and their consumption also poses health risks for the population.

Some people prefer water from wells and rivers to water from pumps because it tastes good. The taste of the water is most often related to the nature of the minerals in the rocks in the region. In Bevougane in the rural commune of Niego, people drink water from the rivers first and when they dry up, they use water from the wells which are also temporary. Water from boreholes is the last resort because it is perennial. In fact, almost all the boreholes in the region are permanent. MARCELIN J., (1971) shows that the optimum depth for drilling in the basement for crystalline rocks is less than 45 m; for shales and green rocks, it varies from 45 to 65 m.

5.4. WATER MANAGEMENT BY USERS
> Within households

Usage practices are not conducive to keeping drinking water safe. Most households use the calabash and the cup as the main tools for collecting water for different needs. According to the results of our surveys, those who do not wash their hands before handling water represent 62% and 24% without soap against 14% who do it with soap. We can say from these figures that people do not take enough hygiene precautions before using water. This behaviour of adults influences the behaviour of children. Children are often allowed to help themselves to drinks. However, they use any kind of container that is readily available to them, regardless of its cleanliness. This behaviour on the part of children increases the risk of contamination of drinking water. This state of affairs can be explained in part by illiteracy. Indeed, 86% of the respondents are illiterate.

Management at the water point

Several manipulations of the water are observed at the sampling site. The water is first collected in a main container. It is then transferred to other secondary containers. When there are many women at the pump, the water is drawn in order of arrival. It is at this point that a mutual aid system is created, which requires the water to be transferred from the main container to another using a calabash or a box. From these manipulations, water can be contaminated from the source. Looking at photo 4 below, we see that there are three contacts with the water before it is transferred to the final transport container. First the basin, then the box and finally the funnel.

Photo 5: Water transfer at the source in Ouessa

Author : MALO Y. 07/01/2013

Water management during transport

The different means of water transport in the region, are reprësentës in the graph below.

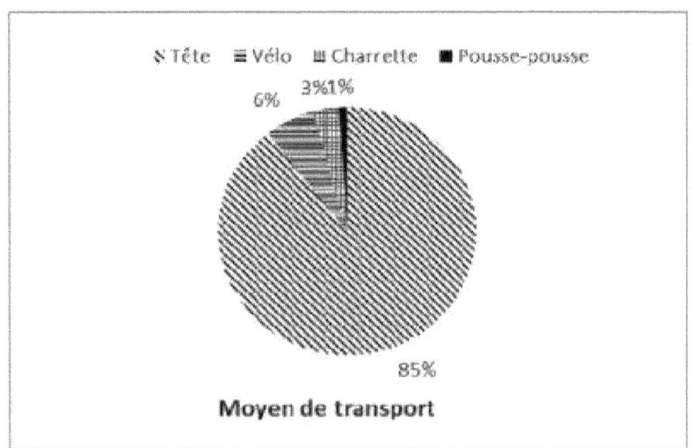

Source: Field survey (December 2012 to January 2013)

Figure 4: Proportions of means of transport

The analysis of this graph shows that feet are by far the most used means of transport and the rickshaw is the least used. Those who use feet when transporting water are the same ones who use basins (73%), as a transport container. The basins are not covered and when the distance is long, women use stabilizers (36% of women who use basins). All these behaviours increase the risk of water contamination, even if it is drinkable at the source. 55% of women use calabashes as stabilisers, followed by plastic (22%), leaves (18%) and textile fabrics (2%). The choice of these different modes of transport could be explained by the relief of Birimien. For example, it is more difficult to use a rickshaw to climb a hillside or a hillock than to do so on foot with a load. Therefore, the volume of the transport container must be small.

5.5. WATER STORAGE

The container and storage time can increase the degree of contamination of the water and the proliferation of bacteria and microbes in it.

> Storage containers

At home, water is transferred to another container for storage. The main ones are: the jar and the barrel. Households using these containers represent 67% and 25% respectively for the jar and the barrel. The jar is used more because it refreshes the water. Women do not wash it with soap. They simply rinse it by hand. Most households, 85%, put their drinking water storage container in the house. But 15% of households do not close it. Cans, which can contribute to a better conservation of drinking water, are used by only 3% of households. These results show that water can lose its potability when stored.

Storage time

This time varies from half a day to more than seven days. People do not store drinking water for a long time, because of its quality (cloudiness). Graph 5 shows the duration of water storage by households.

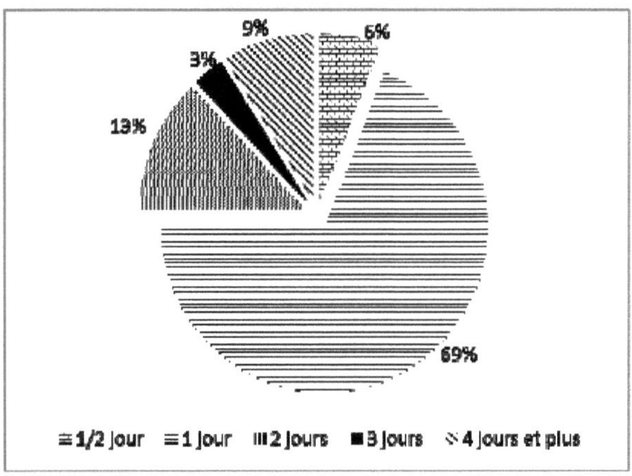

Source: Field survey from December 2012 to January 2013

Figure 5: The duration of water storage

We observe that the majority of households store water for one day (69%), and 9% for four days or more. Looking at Figure 5, we can say that the population in the South West does not store water for a long time. The shorter storage time is explained by the fact that the water they drink is cloudy and contains foreign bodies. After one day, the fillers become visible and the water smells, said a respondent. The women empty this water, which they use either for washing dishes, washing or watering their livestock. In Dadoune in the rural commune of Nidgo, some people prefer to drink the 'dolo' because the water in the wells is cloudy. All these reactions are strategies to cope with the poor quality of water. In Koper, for example, a woman suggested that she only drinks "dolo". For water that tastes of potash, people drink it by adding either tamarind or by adding "to". The latter two methods help them to drink the water more easily, but do not solve the problem of its potability. The soils in the region are chemically rich (see page 22). They influence the taste of the water. They are porous and the water, before feeding the water table, first passes through them. According to NAKOLENDOUSSE S., (1991) water is contained in the rock, one cannot make a study of the content without going through the container insofar as it is the parameters of the latter which condition its existence.

5.6. WATER TREATMENT BY HOUSEHOLDS IN THE SOUTH-WEST

The treatment of raw water depends on the quality, which is a function of its origin and can vary over time. In this study, we were interested in the different methods of water treatment before consumption. From our data, only 12% of households treat their drinking water. Filtering through a cloth and a sieve is the most common type of treatment (50% of those who treat their water), followed by boiling (29%) and other forms such as bleach, tablets (21%). The explanations for this low rate of treatment of drinking water could be related to ignorance, lack of products for some and water deemed clean for others. There is therefore a need for awareness raising and health education. The populations who consider the water to be clean get it from boreholes or standpipes. According to our results, the population is ready (all the households surveyed) to use water treatment products, provided that these products are available and accessible. Water, in its current form of consumption in the South West, has consequences for the health of the population.

5.7. WATER-RELATED DISEASES

Water-related diseases are a serious threat to human health. Many diseases are caused solely by using unsafe water for drinking or cleaning food. Others are caused by inadequate sanitation facilities and poor hygiene habits, linked to the lack of clean water. A distinction is made between water-borne and water-related diseases. Waterborne diseases are those caused by water contaminated with human, animal or chemical waste. They include cholera, typhoid, polio, meningitis, hepatitis A and E, and diarrhoea, among others. Waterborne diseases are transmitted by aquatic organisms that spend part of their lives in water and part as parasites. These diseases are caused by a variety of worms. These worms infect human organisms and are not necessarily fatal (schistosomiasis). And finally, those carried by mosquitoes and tsetse flies that infest certain water areas. These include yellow fever, sleeping sickness, filariasis and malaria, which is the most common.

In the south-west some of these conditions and others that could not be named are present. Thus, according to our surveys, 65% of the households surveyed have problems related to water consumption. The graph below shows the proportions of the different diseases.

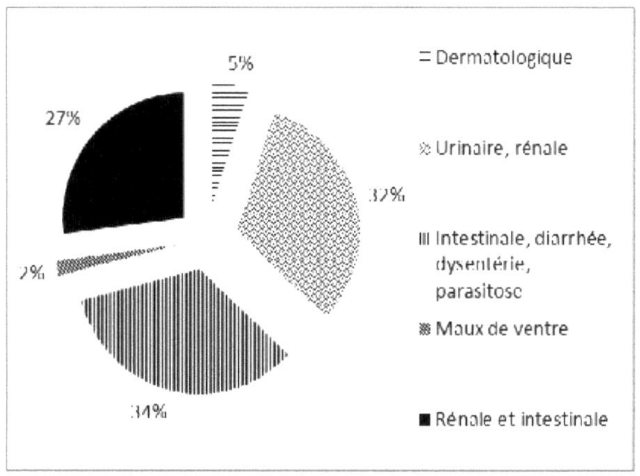

Source: Field survey from December 2012 to January 2013

Figure 6: Waterborne diseases in the South West region

These diseases are caused by drinking and contact with water. Urinary, kidney and intestinal diseases are the most frequent. The same diseases were identified as the main causes of consultation in the health centres. Diarrhoea and malaria are the most commonly diagnosed. Diarrhoea affects more children aged between zero and five years (100% of CSPSs say so). In Ouessa, 70% of patients suffering from diarrhoea are children. This shows that children are more vulnerable to diarrhoeal diseases as a result of unsafe water.

These diseases are common in the region during the months of June, July, August and September. These are the rainiest months and therefore favourable for the proliferation of insects (mosquitoes, flies). Indeed, the climatic aspect is important and is among the risk factors for health. These are also the months when people are most likely to use non-drinking water sources because they are very accessible. Similarly, they spend their time in the field, far from drinking water sources. Bilharzia is particularly observed during these months in children aged 5 to 15 years because of bathing. The level of knowledge of waterborne diseases among the population is noteworthy. According to the analysis of our data, 47% of households do not know that certain diseases are linked to water. According to the health workers, the population does not follow their advice correctly. Some people do not respect the advice and find that water is not responsible for any disease.

CHAPTER 6: HYGIENE AND SANITATION IN THE SOUTHWEST REGION

Sanitation is a mechanism for the collection and disposal of waste produced daily by man through his food chain in order to ensure the quality of his environment and living environment. It goes hand in hand with hygiene, which is also a mechanism that improves the quality of life and health of man through cleanliness of food, body and clothing. For the Ministère de l'Agriculture et de l'Hydraulique, (2011), sanitation and hygiene are determined by the following factors:

- the poverty that affects a large part of the population;
- socio-cultural habits;
- illiteracy and inadequate health education facilities ;
- inadequate and/or inappropriate sanitation systems ;
- non-compliance with environmental legislation ;
- insufficient awareness-raising campaigns.

In addition to the human deficiencies, the birimic environment favours run-off from the peaks into the depressions that constitute the water reserves (natural or man-made). Indeed, these reserves are filled with waste from human and animal dejecta, as well as toxic waste from spraying, cultivation and gold panning sites.

Awareness of the importance of sanitation for human development really emerged in the 1990s, with the International Drinking Water and Sanitation Decade and the World Summit on Environment and Development in Rio in 1992. It was reinforced at the World Summit on Sustainable Development in Johannesburg in 2002. In this context, sanitation was taken into account in Burkina Faso in the 1990s, followed by the elaboration of the first sanitation sub-sector strategy document in 1996 (Monographie nationale sur l'assainissement, 2010).

6.1 AVAILABILITY OF SANITATION SERVICES

Sanitation is an action that aims at the improvement of all conditions in the physical environment of human life that influence or are likely to influence défavorablement on physical, mental or social well-being, DIABATE M., (2010).

1.1.1. Sanitation infrastructure

Latrines

Latrines are sanitation facilities. The Direction Gënërale de l'Assainissement, des Eaux Usëes et Excreta (DGAEUE, 2011), distinguishes:

- the traditional latrine, which is a simple pit covered by a slab with a drainage hole and a cabin for privacy (the cabin for privacy can be made of banco or secco).
- the sanplat latrine: this is a single pit latrine, but it has no ventilation pipe. The drainage

hole is covered with a lid to prevent flies from entering. The pit can be square or circular. The concrete slab can be dome-shaped or flat. This type of toilet, when improved, has an air pipe and a fly screen.

- the ventilated pit latrine is a structure consisting of a set of slabs, a cabin, and a ventilation pipe per pit, the end of which is fitted with a fly screen. It includes one or more sludge reception and accumulation pits with constructed walls.

- the ecosan latrine (Ecological Sanitation). It is a fully above-ground latrine built on a concrete platform. The slab has two drainage holes, footrests and a slope in the slab for the separation of urine from feces. In the front there are one or two compartments, with a canister and a ventilation pipe.

- the mechanical flush toilet: this consists of a water tank which supplies water to evacuate secretions. It is equipped with a Turkish or English-style bowl, constituting the siège of defecation.

Photo 6: Example of a traditional latrine

Author : MALO Y. 06/01/2013

Public latrines

It is difficult to speak of public latrines in the South West region. In all the villages visited during this study, public latrines are only found in public services such as CSPSs and primary schools. Access to these latrines is limited to patients, their carers and nurses in the case of the CSPS. As for the latrines in public primary schools, their use is reserved for schoolchildren and their teachers. Markets and other places where large numbers of people can gather do not have them.

Family latrines

Family latrines are also not numerous. Only 21% of the households surveyed have one. However, this situation hides disparities. In Noumbiel and Poni, no household surveyed had a latrine. The majority of households with latrines are in Iola (95%). The remaining 5% of households with latrines are in Bougouriba. Throughout the region, there is a lack of latrines and those that do exist do not meet national standards. According to the DGAEUE, a household has access to family sanitation if it uses an improved latrine (whether shared with other households or not) with a total number of daily users of ten or less. These are VIP, Ecosan, Manual Flush Toilet (MFT) and Mechanical Flush Toilet. In the region, the households surveyed do not have any and those available are traditional and in poor condition. Map 12 shows the spatial distribution of latrines in the region.

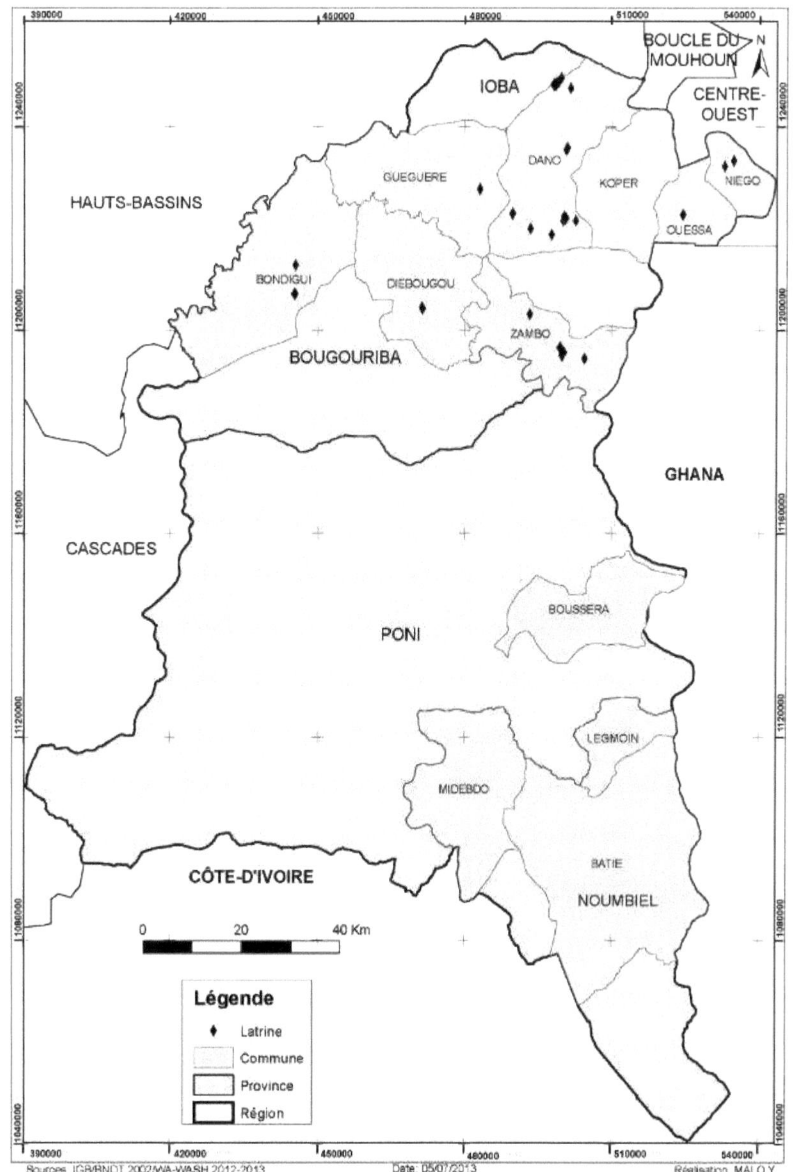

Map 12: Distribution of traditional latrines

Bins and gutters

Gutters and dustbins are rare in the South West. Gutters are found in the urban communes of Dano, Diëbougou and Gaoua. Their construction is made possible by the asphalting of roads. In rural areas, gutters are non-existent. There is also a shortage of rubbish bins in the villages.

1.1.2. Distances from rubbish dumps to water sources

In the south-west, apart from pollution from agricultural inputs and gold mining sites, another source is the waste produced by households. The waste produced is deposited at varying distances from water sources. Our surveys show that 15% of households have their rubbish dump within 100 metres of their water source, 27% within 500 metres and 58% of households more than 500 metres away. In doing so, 15% of households rely on water sources that are at risk of being polluted. These households have their water source close to their concession or within their course. Most of the time, these are traditional wells that have no coping. At the slightest rain, the water is drained away with rubbish. Photograph 6 below shows a traditional well in Balembar.

Photo 7: Traditional well

Author : MALO Y. (20/12/2012)

1.1.3. Sanitation at the family level

The family sanitation access rate is the number of households with access to family sanitation relative to the total number of households (NP-FSAP, 2006). Based on this rate, the table below is obtained.

Table 12: Sanitation access rates by province

Provinces	Number of households surveyed	Number of households with latrines	Sanitation access rate
Bougouriba	265		19%

loba	155	58	37%
Noumbiel	14	0	0%
Poni	5	0	0%

Source: Field survey from December 2012 to January 2013

The analysis of the table shows that the rate of access to sanitation is almost zero for the provinces of Noumbiel and Poni. However, it is 19% for Bougouriba and 37% for Loba. The access rates of the first two provinces (Nombiel and Poni) are below the national rate of the PN-AEPA (2011), which is 10%, and that of the Joint Monitoring Program (JMP), which is 6%, in 2011. The rates in the other two provinces (Bougouriba and Ioba) are above the PN-AEPA rate and the JMP rate. Overall, these rates are below the MDG target of 54% access to sanitation in 2015. Not all of these latrines are emptied before our visit. This suggests that they are either recent or not frequently used by the inhabitants of the households. The relief and lithology of the Birimien can make it difficult to build certain infrastructures, particularly those for sanitation (gutters, latrines, cesspools), in the region.

6.2. SANITATION PRACTICES

With the lack of latrines, nature remains the only place to go. The same applies to wastewater management when there are no cesspools or septic tanks. A survey conducted in 2009 by the DGAEUE shows that in Burkina Faso, defecation in the wild concerns six out of 10 households and in rural areas, it represents eight out of 10 households.

6.2.1. Nature as a place of degradation due to lack of sanitation infrastructure

For the vast majority of people in the South-West, nature remains the only place of sanitation. In households with latrines, we observe that not all family members use them. Some family members prefer to use the bush, despite the existence of a latrine. In Bontioli in the rural commune of Zambo, for example, an enquëtë made it clear that he ëlɔк the only one who does his business in the family latrine. The other family members prefer to use the bush for their needs. This situation is very common in the region. Thus, the analysis of our data shows that 61% of the respondents defecate in the wild. There are disparities from one province to another as shown in Figure 7 below, which presents the situation by province.

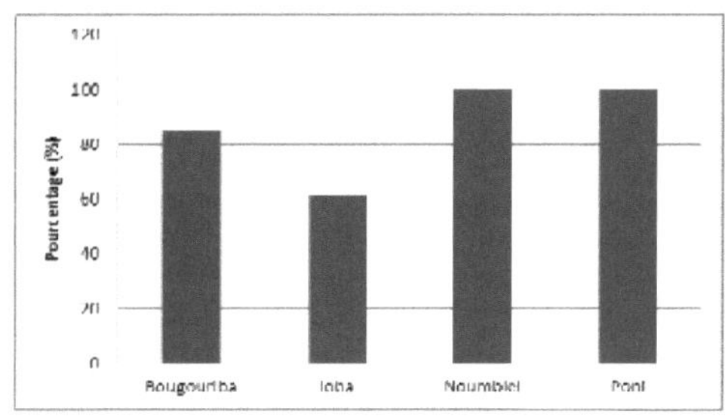

Source: Field survey from December 2012 to January 2013

Figure 7: Populations defeating in the wild

The graph shows that in the province of Loba, 61% of respondents dëfës in the wild, 85% in the province of Bougouriba and 100% in the provinces of Noumbiel and Poni. These results are above the national average of 60% of households that dëfëquent in the wild (DGAEUE, 2010). It can therefore be said that there is access ПпШë to sanitation in the South West.

6.2.2. Waste management by households

Waste is variously managed by mënages. These are solid and liquid dëchets. Organic waste that is completely biodegradable is increasingly giving way to a complex hëtërogëne of more or less biodegradable waste.

> Management of household waste

Following our observations, the majors of the mënages investigated swept aside their concession. This is something to be appreciated. But, how is this rubbish disposed of? The disposal of household waste is an important aspect of a healthy living environment. There are several ways in which the rubbish is disposed of in the Southwest. These are disposal by incineration, by use in pits to produce organic manure, by their dumping in rubbish heaps, in the street and in the fields. We translate these different modes of disposal in Figure 9 below. It shows that 45% of households put their waste in manure pits, 32% in rubbish heaps, 11% in the street, 4% incinerate it and 4% in the hut fields, 3% have no fixed place to throw it away and 1% throw it in cesspits.

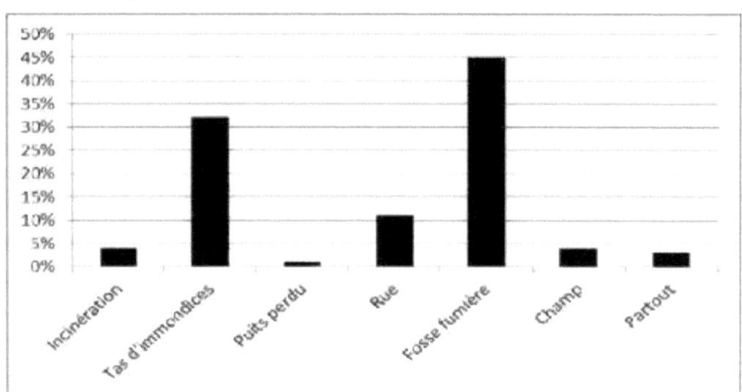

Source: Field survey from December 2012 to January 2013
Figure 8: Methods of waste disposal.

The most acceptable disposal is in manure pits. It exposes the rubbish less to wind and rainwater run-off. In addition, the waste is turned into fertilizer for the fields. This is the primary purpose of the manure pit. On the other hand, they can in some cases turn out to be nests for mosquitoes and therefore a source of contamination due to their proximity to the concessions. The majority of households (55%) expose their rubbish to winds and streams which can carry it to rivers. This solid waste is one of the most significant environmental hazards, polluting water, soil and air. The slopes of the Birimien also make it easy to carry it into the rivers. In fact, along the streams or marshes, rubbish is piled up, causing water, soil and air pollution.

Wastewater management

The situation of wastewater disposal is not so much better in the South West region. This is domestic water generated by cooking, washing up, laundry and showers.

Shower water: Our observations revealed that most showers discharge their water into the environment. There is no cesspool or septic tank. This stagnant water is a nest for mosquitoes. This wastewater is mainly composed of phosphate, carbon, nitrogen and many pathogenic germs. And according to WHO (1996), the absorption of such water in any way can be harmful to health in the short or long term. In the region, in one way or another, these substances can enter the food chain of the population. Photograph 7 shows the water from a freshwater plant being discharged into the environment.

Photo 8: Discharge of water from a shower into the wild in Ouessa.

Author : MALO Y. 07/01/2013

Water from dishes, kitchens and laundry: several places are used by households to discharge this water. They are discharged into the compound, into cesspools, into the street, into manure pits, into water troughs and into the environment. Manure pits and cesspools are the most acceptable modes of disposal and are used by 32% and 2% of households respectively in the South West. Figure 9 shows the different modes of wastewater disposal.

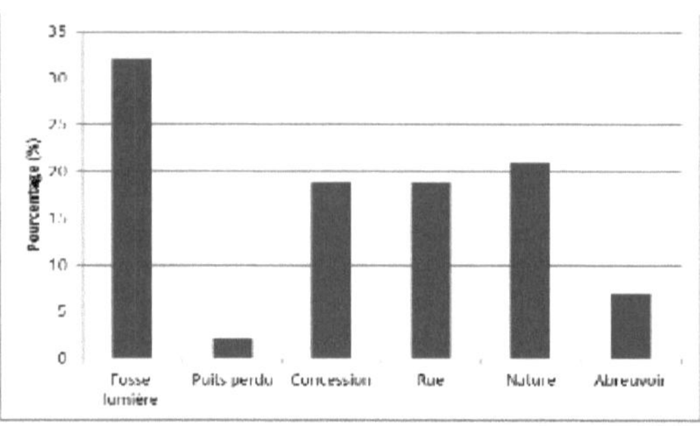

Source: Field survey from December 2012 to January 2013
Figure 9: Method of wastewater disposal

Households who dispose of wastewater in this way are often unaware of the health risks associated with their proximity. Indeed, 70% of the respondents were aware that these disposal methods can cause disease. Of these, 33% could not identify a single disease associated with these methods. On the other hand, 5% of the respondents felt that these disposal methods could not cause illness and 25% did not comment on the issue. This leads us to say that people are not aware that some diseases are linked to bad wastewater disposal. Liquid waste is a crucial issue for the well-being of the environment. The disposal of excreta and urine in unsanitary conditions is one of the most common causes of environmental contamination. The stagnation and infiltration of wastewater are factors that influence the degradation of environmental elements, especially groundwater.

6.2.3. Health education

It is done through awareness campaigns on the one hand and by health workers in the CSPSs on the other. The level of knowledge of diseases related to water, hygiene and sanitation shows that the population is unaware of the health risks that this could cause. Today, Plan Burkina's representation in Gaoua, with the support of the NGO Eau et Assainissement pour l'Afrique (EAA), is implementing health education modules in primary schools. To this end, teachers have received training in hygiene. Hand-washing devices have been installed in schools and drinking water bags have been installed in classrooms. This health education only concerns the provinces of Noumbiel and Poni. In these provinces, it only concerns schoolchildren, with the hope that they can influence the behaviour of those around them. However, in Ioba, particularly in the village of Tambiri, a schoolboy tells us that when he tried to sensitise his parents about hygiene practices, he was violently reprimanded. This tells us that this Plan work will have an effect, but in the long term. There is also health education by health workers. It has a low impact since it is directed towards the patients. Indeed, health workers raise awareness during vaccination campaigns and during consultation sëances. The lack of means means that there is no follow-up and in the majority of cases, their advice is not respected by the patients.

6.3. HYGIENE PRACTICES

In our study, we were interested in hygiene before meals and after bowel movements.

6.3.1. Hand washing with soap before meals

Many people do not wash their hands with soap before meals. In the South West region, our work shows that 18% of the respondents wash their hands with soap before meals, compared to 82% in the rest of the country. A great deal of work remains to be done to ensure that all of these populations integrate this into their daily practice. In the villages, most families eat their meals together. To do this, they all wash their hands in the same water. The oldest usually washes his hands first, leaving some germs and picking up others, and the next ones leave some germs and pick up others in the same water. If one has not washed one's hands with soap before the meal, one will have the same attitude afterwards. In this case, one is a vector of germ transmission when greeting others. A different situation is observed in public places.

If families at least wash their hands without soap before meals, in markets and other gathering places, another situation arises. People eat boiled meat at the market, pancakes and doughnuts. They eat them without washing their hands. In this case, they run health risks because the meat is not protected against dust and, worse, the hand that removes it is probably soiled after having shaken several other hands. In the butcher's shop, the same hand serves the meat and collects the money at the same time.

6.3.2. Hand washing after bowel movements

In all of the households reached in this study, handwashing after bowel movements is rare or non-existent. People do not take care to wash their hands after bowel movements. Our surveys showed that 25% wash their hands with soap after defecating, 34% without soap and 45% not at all. This can be explained by the fact that the majority of the above-mentioned households (61%) defecate in the open. It should be remembered that 86% of the respondents are illiterate.

Food hygiene is not to be neglected. For the provinces of Noumbiel, Bougouriba and Poni, the different health districts did not provide any information, but in the health district of Dano, cases of intoxication and poisoning were recorded during the year 2012. According to the Chief Medical Officer of the Dano health district, these cases of intoxication are mainly observed among dolo drinkers.

6.4. IMPACTS ON THE ENVIRONMENT AND HEALTH OF THE SOUTH-WESTERN POPULATION

6.4.1. The impact on the environment

The environment is the substratum and all that it harbours, i.e. water, air, all the physical, chemical or biological elements of nature (DIABATE M., 2010). In the South-West, the lack of sanitation infrastructure, the lack of health and environmental education as well as inappropriate sanitation practices are harmful to the environment. Indeed, the region's environment is an open-air bin. It is the place where people are defecated. The soiling of our living environment and the non-respect of hygiene practices have harmful effects on the health of the population.

6.4.2. The impact on the health of the population

Residual waste constitutes a very important nuisance environment (presence of bacteriological loads), and serves as a breeding ground or survival site for numerous microbial species, responsible for various endemic and epidemic diseases (cholera, diarrhoea, malaria, dysentery). Worms (roundworms, whipworms) acquire their contaminating power over a more or less long period of time depending on the quality of the external environment. Mosquitoes, pathogenic agents of malaria, proliferate in an unhealthy environment. Flies also thrive in an unhealthy environment and their mode of contamination can be direct or indirect (e.g. by landing on tax materials and then transmitting the infectious agents by landing on our food). In this way, humans contract life-threatening diseases such as typhoid fever. In Burkina Faso, these diseases are the leading causes of consultation. In the South-West region, all the CSPSs visited as part of this study found that diseases related to hygiene and sanitation were the most frequently diagnosed. These are malaria, diarrhoea, dysentery, parasitosis; anthrax and some cases of poisoning are reported (191 cases in the Dano health district for the year 2011-2012). Our surveys reveal that 78% of households have children under five years of age and 45% of them had diarrhoea during the last two weeks of our visit. In 2009, nationally, diarrhoea accounted for 70% of the causes of consultations for children aged between zero and five years. And only 5% of women have knowledge of diarrhoea and its treatment (DGAEUE, 2009).

PARTIAL CONCLUSION

Burkina Faso is in a spiral of poverty, the collateral consequences of which include lack of access to drinking water, sanitation and hygiene. In the South-West region, other factors are added to this. Indeed, the physical environment, in particular the rock structure and the thickness of the soil, are decisive. The rapid drainage of rainwater due to the high relief, the poor retention of this water by a less thick soil, prevent a good recharge of the water tables. They are one of the main sources of water supply for the populations of the region. In addition to these physical factors, the behaviour of the populations of this region of the country does not guarantee the availability of the water resource on the one hand, and on the other hand, compromises the potability of this water, even if it is at source. These behaviours are justified by the poor practices adopted by the populations of the South West in terms of drinking water, hygiene and sanitation. These factors impact on access to drinking water, hygiene and sanitation. However, disparities exist between the rate of access to sanitation and hygiene, which appears to be lower than that of access to drinking water. They also have implications for the health of the population in the South West.

GENERAL CONCLUSION

The study on drinking water, hygiene and sanitation in the South-West region was an opportunity for us to get a feel for the living conditions of households in this part of Burkina Faso. It allowed us to highlight a number of problems related to this topic.

The water points used in the region are: boreholes, traditional and modern wells and standpipes. In addition, there are the primary water points that are surface waters. A total of 707 water points have been identified with the help of the GIS tool.

The mapping has enabled a spatial distribution of these water points in the region. It can be seen that some communes are better off than others in terms of availability of drinking water. The level of drinking water coverage is less than 50% for some communes and even zero at the village level. These water points represent 28% of the water points in the region. In addition, there is an under-equipment of infrastructure. This is exacerbated by the breakdowns that put a good part of these infrastructures out of service (40 boreholes out of 141 are not functional and eight standpipes out of 16 are not functional). As for non-drinking water points, they represent 72% and are unevenly distributed from one province to another, from one commune to another and from one village to another. There is therefore an unequal distribution and a lack of drinking water infrastructure. The physical environment, i.e. the relief of the Birimien, influences the unavailability and inaccessibility of drinking water through the depth of the water table and therefore the quantity available. We can therefore say that our first hypothesis on the unavailability of water linked to its unequal distribution in the South West region is confirmed.

In the southwest, drinking water sources are not close to households. This influences the frequency of use of drinking water points. In addition, the relief of the Birimian region does not facilitate the transport of water in the region, so it is difficult to bring a large quantity of water at once. Thus, the average time to collect water is 3 hours per day. The price of water creates exclusion. Some households do not have access to drinking water, simply because they have not paid their subscription. Even if this population has access to drinking water, the usage practices make it lose its potability. Our second hypothesis on the inaccessibility of drinking water is thus confirmed.

As for hygiene and sanitation, infrastructure and practices are not better. Public latrines are rare and are mostly found in public institutions. Family latrines are used by only 21% of households in the South West. The rate of access to sanitation is ini'erieur at 50%. Nature remains the main place of sanitation. All these behaviours increase the prevalence of certain diseases. Hygiene practices are not good either and most households do not take enough hygiene precautions. Given the level of infrastructure and the sanitation and hygiene practices, we say that our third hypothesis on diseases related to lack of hygiene and sanitation is true.

Synergy of action is needed in this case. It must involve all actors (beneficiaries in the first place, decision-makers and civil society (NGOs and associations)). For a better intervention, we will suggest some recommendations for the different stakeholders in the South West region, in the WASH sector. In order to help local communities to address this problem, it would be necessary to :

- increase the number of sustainable WASH facilities;
- Establish viable water users' associations through the involvement of beneficiaries;
- Continuously raise awareness of the use and maintenance of the structures;
- build water reservoirs to support groundwater and agriculture;
- for hygiene, create permanent awareness;
- for sanitation, be rigorous about respecting environmental standards, which will lead people to adopt good practices. But this requires environmental education in the region and
- to raise awareness of the need to clean drinking water. The use of Aquatab tablets would be desirable for this purpose.

BIBLIOGRAPHY

1. General works

1) ARNOULD M., (1976). Etude Gëologique des Migmatiques et des Granites Precambriens du nord-est de la Republique de la Cote d'Ivoire et de la Haute-Volta mëridionale. *These, Memoires, B.R.G.M, N°3, Paris,* 174 pages.

2) BALANCA G. et *al,* (2007). Birds of the WAP Complex, 199 pages.

3) World Bank (WB), (2007). Water and Sanitation Program (WSP-FA), 66 pages.

4) BRUNO V. and MARINA R., (2012). Le Livre Bleu S/C de L'Eau Vive Burkina. *Ouagadougou,* 36 pages.

5) Bureau de Recherches Geologiques et Miniëres (BRGM)/Aquater, 1985. Etude de la Recherche par les Pluies des Aquiferes des Milieux Fissures: *Rapport d'Avancement n°2, Ouagadougou,* 27 pages.

6) CISSE G., and TANNER M., (1999). Suivi et Evaluation des Risques Sanitaires Lies a l'Utilisation des Eaux Polluees en Agriculture Urbaine a Nouakchott (Mauritanie) et Ouagadougou (Burkina Faso), 13 pages.

7) Comite Inter Africain d'Etudes Hydrologiques, (1989). Contribution a la Methode de Prospection des Eaux Souterraines sur le Bouclier Cristallin d'Afrique de l'Ouest: Etudes Hydrologiques et Geophysiques au Burkina Faso. *Ouagadougou CIEH*, 168 pages.

8) Delegation of the European Commission, (2006), Final Report: Environmental Profile of Burkina Faso, 66 pages.

9) Deuxieme Conference des Cadres, (1982). Ministere du Developpement Rural (Haute Volta) *Annexe III : Secteur Eau et Developpement. Ouagadougou,* 63 pages.

10) DEZTTER A., (1996). Les Enjeux de la Gestion des Ressources en Eau en Milieu Semi-Aride. *Ouagadougou, ORSTOM,* 25 pages.

11) Direction Generale de l'Amenagement du Territoire, du Developpement Local et Regional/Ministere de l'Economie et des Finances, 2010. Profil des Regions du Burkina Faso, 456 pages.

12) Direction Gënërale de 1 Assainissement des Eaux Usëes et Excreta (DGAEUE), (2010). Enquete Nationale sur les Conditions de Vie des Mënages aux Ouvrages d'Assainissement Familial (Monographie Rëgionale Sud-ouest). *Ouagadougou*, 64 pages.

13) Direction Gënërale des Ressources en Eau (DGRE), (2013). National Confërence on Public-Private Partnerships (PPP) in the field of drinking water and sanitation.

14) Direction de l'Inventaire des Ressources Hydrauliques, (1993). Carte Hydrogëologique du Burkina Faso au 1/500000ᵉ Feuille de Gaoua. *Ministere de I'Eau du Burkina Faso,* 37 pages.

15) FONTES J. and GUINKO S., (1995). Notice Explicative de la Carte de la Vëgëtation et de l'Occupation des Sols du Burkina Faso. *Universite de Ouagadougou*, 71 pages.

16) GENDREAU F., GUBRY P. and VERON J., (1996). Population et Environnement dans les Pays du Sud. *Paris, Karthala (CEPED)*, 309 pages.

17) HUGOT G., (2002). A la Recherche du Gondwana Perdu : aux Origines du Monde (U.M.R) du C.N.R.S. ("Espace") 6012, 309pages.

18) Institut National de la Statistique et de la Demographie (INSD), (2009). Monographie de la Rëgion du Sud-Ouest. *Ouagadougou*, 152 pages.

19) Institute of Research for Development (IRD), (2010). Activity Report (2010), 66 pages.

20) KY C. J. and GOMGNIBOU M., (2008). Rapport de Recherche de l'Equipe de Collecte de Traditions Orales sur les Ruines de Loropëni (Annex 1). *Ouagadougou*, 91 pages.

21) LAMARQUE F., (2004). Les Grands Mammifëres du Complexe WAP, 51 "page" cards.

22) MARCELIN J., (1971). Notice Explicative de la Carte Gëologique au $1/200000^e$ de la feuille de Gaoua - Batië. *Paris, edition: BRGM (Bureau de Recherches Geologiques et Minieres)*, 29 pages.

23) Ministëre de l'Agriculture, de l'Hydraulique et des Ressources Halieutiques, (2008). Capitalisation des Bonnes Pratiques et Technologiques en Agriculture Imguee. *Ouagadougou, UNDP*, 165 pages.

24) Ministëre de l'Economie et du Developpement, (2005). Cadre Strategique Regional de Lutte Contre la Pauvrete. *Ouagadougou*, 58 pages.

25) Ministry of Water and Environment, (1998). Politique et Strategies en Matiere d'Eau, 119 pages.

26) Ministry of Environment and Living Environment, (2007). Programme d'Action National d'Adaptation aux Changements Climatiques (PANA). *Ouagadougou*, 71 pages.

27) Ministry of Environment and Water, (2001). Etat des Lieux des ressources en Eau du Burkina Faso et de leur Cadre de Gestion, 252 pages.

28) MONJOUR L.; KABORE A. and BONKOUNGOU D. R., (1997). Rapport Final d'Activite. Comite de Jeune pour l'Eau Potable, l'Hygiene et l'Assainissement dans l'Arrondissement de SIG-NONGHIN dans la Ville de Ouagadougou, 27 pages.

29) PERRE M., (1982). Les deux bouches : les societes des " Rameaux Lobi " entre la tradition et le changement, Tom I. Universite de Pantheon-Sorbonne; Paris, 701pages.

30) Rouville (de) C., 1987. Social Organization of the Lobi of Burkina Faso and the Ivory Coast. *Paris, L'Harmattan*, 259 pages.

31) SANTOS D., (2005). Koom la viim: Enjeux Socio-Sanitaires de la Quete de l'Eau a Ouagadougou. *Montreal, UM*, 221 pages.

32) SAVADOGO A. N., (2012). Champs Captant sur Mega-fractures du socle et Generalisation des Reseaux d'Adduction d'Eau Potable en Milieu Rural au Burkina Faso. *Universite de Ouagadougou : Laboratoire d'Hydrogeologie*, 12 pages.

33) ULLMANN C., (2012). Rapport d'Ingenieur : Alimentation en Eau Potable des Populations en Milieu Rural, Contamination des eaux Souterraines par l'Arsenic. *ANTEA. BF*, 147 pages.

34) <u>WaterAid and VARENA-ASSO,</u> (2011). Plan Communal de Developpement Sectoriel Approvisionnement en Eau Potable et Assainissement de Dano, 77 pages.

2. Theses and dissertations

35) ABOU M., (2010). Contribution du Systeme d'Information Geographique a la Planification des Infrastructures en Eau Potable dans la Commune Rurale de Tanghin-Dassouri, Region du Centre. Ouagadougou, Memoire de Master, UO/ Departement de Geographie, 104 pages.

36) BAKO F., (2008). Suivi Spatio-temporel des Ressources en Eau des Reservoirs Artificiels et de leur Usage. *Ouagadougou, UO/Departement de Geographie*, 39 pages.

37) DA D. E. C. F., (1980). Contribution a 1'Etude Gëographique des Paysages Voltaiques, Monographie de la Rëgion de Gaoua. Tome I, 151 pages.

38) DIABATE M., (2010). Dëchets Mënagers : Impact sur la Sante et 1'Environnement en Commune I du District de Bamako : cas de Banconi. *Bamako, Memoire de Maitrise, (web source 25/01/2013).*

39) GALBANI S. R. P., (2011). Strategic' de Gestion des Points d'Eau Potable dans la Commune de Ziniare. *Master's thesis, UO*, 86 pages.

40) ILBOUDO K. I., (2008). Impact du Barrage de Loumbila sur la Sante des Populations Riveraines de Pousghin. Ouagadougou, *Memoire de Maitrise, UO/Departement de Geographie*, 103 pages.

41) KABORE O., (2008). Dynamique des Plans d'Eau Liee a la Variabilité Climatique et a l'Action de l'Homme. *Ouagadougou, Memoire de Master, UO/Departement de Geographie*, 112 pages.

42) KABORE S., (2008). Acces a l'Eau Potable et Assainissement: Effet sur la Morbidite Diarrheique des Enfants a Ziniare. *Ouagadougou, Memoire de Maitrise, Departement de Geographie*, 90 pages.

43) KAFANDO Y., (2004). Environnement Urbain et Probleme de Sante a Ouagadougou : cas du Quartier Cissin. *Ouagadougou, Memoire de Master, UO/Departement de Geographie*, 112 pages.

44) KAGUEMBEGA I. P. F., (1999). Acces aux Soins de Sante en Amont du barrage de

Bagre. *Ouagadougou, Master's thesis, UO/department of Geography*, 112 pages.

45) KEITA D. M. W., (2010). Contribution du Programme Saaga a l'Amelioration des Conditions Pluviometriques dans le Bassin du Nakambe. *Ouagadougou, UO/Departement de Geographie*, 89 pages.

46) KIMA A., (2006). La gestion des Dechets Solides Menagers a Koudougou. *UO, Memoire de Maitrise, Departement de Geographie*, 147 pages.

47) KYELEM C. M., (2009). GIS and Method of Spatial Distribution of Primary Schools in the Amenagee Zone: Case of the Bogodogo District of the City of Ouagadougou. *Master's thesis, UO/Departement of Geography*, 100 pages.

48) NAKOLENDOUSSE S., (1991). Methode d'Evaluation de la Productive des Sites Aquiferes au Burkina Faso. *Grenoble, These, Universite Joseph Fournier, 258 pages.*

49) OBOULBIGA S. R., (2008). Acces a l'Eau Potable sur les Rives du Lac Bagre. *Ouagadougou, Memoire de Maitrise, UO/Departement de Geographie*, 98 pages.

50) OUANGRE J. V. T., (2009). Analyse des Risques Sanitaires Lies a l'Eau a Ouagadougou : Cas du Quartier Tanghin dans l'Arrondissement Nongre-Masson. *Ouagadougou, Memoire de Masrer, UO/Departement de Gegraphie*, 69 pages.

51) OUEDRAOGO M., (1993). Le Role du Barrage de Tanghin Wobodo dans la Lutte Contre l'Insecurite Alimentaire. *Ouagadougou, Memoire de Maitrise, UO/Departement de Geographie*, 119 pages.

52) SAVADOGO M., (2009), l'Acces a l'Eau Potable et l'Assainissement dans les Peripheries de Ouagadougou : Cas de la Trame d'Accueil du Secteur 30. *Memoire de Maitrise, University of Ouagadougou*, 105 pages.

53) TRAORE G. H., (2009). Atouts et Contraintes de l'Exploitation des Eaux Useses en Agriculture Urbaine : Cas de Kossodo et Wayalghin. Ouagadougou, *Memoire de Maitrise, UO/Departement de Geographie*, 114 pages.

54) YAO T. M., (2005). Systeme d'Information Geographique Lies aux Dechets Autour des Marches: Cas des Secteurs 25 et 26 de Ouagadougou. *Ouagadougou, Memoire de Maitrise, UO/Departement de Geographie*, 110 pages.

55) YONI L. T. B., (2010). La Distribution de l'Eau Potable dans la Ville de Ouagadougou : Mise a Journee des Données Spatiales de Base. *Ouagadougou, UO/Departement de Geographie*, 97 pages.

3. Reviews and publications

56) INFO CREPA, (2004). Quarterly of the Centre Regional pour l'Eau Potable et l'Assainissement a faible cout. *N°43, March 2004, Ouagadougou, Burkina Faso*, 30 pages. L'Evenement n°250 of 25 February 2013.

57) Sud Science et Technologie n°2, (1998). Gestion des Ressources en Eau: le logiciel hydram. *July 1998, Ouagadougou, Burkina Faso,* 63 pages.

58) Sud Science et Technologie n°5, March 2000. Ouagadougou, Burkina Faso, 63 pages.

4.   The websites -en.wikipedia.org/wiki/Drinking-water, (12/06/2012);
-http://ethesis.unifr.ch/Ky Abraham.pdf date accessed 25/06/2012 ;
-http:// upload.wikipedia.org date accessed 25/06/2012 ;
-http://hydrologie.org/ACT/ORSTOMXII/V consultation date 25/06/2012.
-Ipsinternational.org/en/ consultation date 06/05/2013 ;
-Eauburkina.org date of consultation 31/07/2013 ;
-Wikipëdia.org/wiki/Burkina Faso 24/09/2010.

ANNEXES

Annex: Household Questionnaire

Identification

N°	Questions and Filters	Codes
001	Questionnaire number	LJ_J_]
002	Region	[_]_]
003	Province	[_]_]
004	Municipality	[_]_]
005	Village	[11]
006	Neighbourhood	
007	Household number	[_I_]
008	Location of the household	X: Y:
009	Survey line number in the household	[_I_]

* Result Codes
1=Completed questionnaire
2=Refusal
3=Different
4=Absent survey
5 = Partially filled

Start time
Investigator Date: //Signature
Supervisor   Date :   L   L   Signature

I. Socio-demographic characteristics of the respondent(e)

N°	Questions	Answers		Instruction
Q101	Name of the survey			
Q102	Gender	Female	0	
		Male	1	
Q103	Age	Years gone by / / / years		
Q104	Status of the survey	Female head of household	1	

N°	Questions	Answers		Instruction
		Male head of household	2	
		Adult female but not head of household	3	
		Adult male but not head of household	4	
Q105	Did you go to school?	No>	0>	If 0! Q107
		Yes	1	
Q106	What level of education did you attend?	Primary	1	
		Secondary	2	
		Superior	3	
Q107	Have you attended a literacy centre?	No	0	
		Yes	1	
Q108	What level have you reached?	Alpha Initial	1	
		Alpha Complementary Basic Specialised Technical	2	
			3	
		Training		
Q108	Household size			

II-WATER

N°	Questions	Answers		Instruction
Q201	Where does the drinking water used by the family come from?	Surface water	1	
		Well	2>	
		Drilling	3	If 2! Q203
		Water system	4>	If 4! Q204
Q202	What is the type of surface water?	Watercourse	1	
		Lake	2	
		Mare	3	
		Dam		

N°	Questions	Answers	Yes	No	Instruction
Q203	What is the type of well?	Traditional	1	0	
		Modern	1	0	
		Protects	1	0	
		Unprotected	1	0	

N°	Questions	Answers		Instruction
Q204	What is the type of water system?	Fountain bollard		
		Individual connection	1	
Q205	Is there an improved water point available? that the family could use?	Yes	1	
		No	2	
		Don't know	3	
Q206	Why don't you use the of water improved ?	Too expensive	1	
		Too far away	2	
		Long waiting lines	3	
		Seasonal stables	4	
		Rationing	5	
		Political reasons	6	
Q207	Does the water system supply enough water? of water all year round?	Yes	1	If 1! Q211
		No	2	
		Don't know	3	
Q208	Why the system does not provide enough of water all year round?	Long periods of unavailability	1	
		Irregular supply	2	
		Low water pressure	3	
		Political reasons	4	

N°	Questions	Answers		Instruction
		Payment problems	5	
		Tap disconnected	6	If 8! Q209

N°	Questions	Answers		Instruction
		Rationing	7	
		Seasonal stables	**8>**	
		Don't know	9	
Q209	In which months do these shortages occur?		Yes	No	
		January	1	0
		February	1	0
		March	1	0
		April	1	0
		May	1	0
		June	1	0
		July	1	0
		August	1	0
		September	1	0
		October	1	0
		November	1	0
		December	1	0
Q210	How many hours a day can you collect water?			
Q211	How far do you travel to fetch water?	-500m	1	
		1km	2	
		+1km	3	
Q212	Do you pay a tariff or user fee for water?	Yes	1	
		No	**0>**	If 0! Q215
Q213	How often do you pay user fees?	Per unit (litre - bucket - canister)	1	
			2	
		Per day	3	
		Per week	4	
			5	
		Per month	6	
		By year	7	
			8	
		By season		
		Package plus unit price Don't know		
Q214	What is the rate/amount of the user fee?			
Q215	What equipment do you use to draw water?	bucket sump	1	
			2	
		Can	3	
		Manual pump	**4>**	If 4! Q220
Q216	Where is the water container stored when not in use?	Remains in the well/workplace Outside - on the ground In a public place In someone's home	1	
			2	
			3	
			4	
Q217	Is the equipment used to draw water cleaned before being immersed in water?	Yes	1	
		No	0	
Q218	Is the same container used for drawing and transporting water?	**Yes**	**1>**	If 1! Q221
		No	0	
Q219	When transferring water, the	Yes	1	

N°	Questions	Answers		Instruction
	Do the containers for drawing and transporting come into contact?	No	0	
Q220	What type of container is used to transport water?	Can	1	
		Jarre	2	
		Bucket	3	
		Basin	4	
		Barrel	5	
Q221	How is water transported from the collection point to the household?	On the head	1	
		By hand	2	
		Charette	3	
		Rickshaw	4	
		Two wheels (bicycle and motorbike)	5	
Q222	Do you use stabilisers during transport?	Yes	1	
		No	**0>**	If 0! Q224
Q223	Which ones?	Plastic Foliage Fabric	1	
			2	
		Other	3	
			4	
Q224	Is the same container used for transport and storage of water?	**Yes**	**1>**	If 1! Q227
		No	0	
Q225	When transferring water, do the transport and storage containers come into contact?	Yes	1	
		No	0	
Q226	43. In what type of container is the water stored?	Can	1	
		Jarre	2	
		Jerrican	3	
		Basin	4	
		Barrel	5	
Q227	44. Do you treat your drinking water?	Yes	1	
		No	**0>**	If 2! Q229
Q228	How?	Boiling Filtration	1	
			2	
		Chlorination	3	
		Tablets Other	4	
			5	

N°	Questions	Answers		Instruction
Q229	Why?			
Q230	Is the storage container open to the air?	Yes	1>	If 1! Q232
		No	0	
Q231	What separates the recipient's water from the open air?	Linen stopper Lid	1 2 3 4	
		Opening by tap		
Q232	Where is the storage container located?	Floor installation - outside	1	
		Floor installation - indoors	2	
		High installation - indoors	3	
Q233	Do you regularly clean the water transport containers?	Yes	1	
		No	0>	If 0! Q235
N°	**Questions**	**Answers**		**Instruction**
Q234	At what frequency?	Every week	1	
		Every 2-3 days	2	
		Every day	3	
		Before filling each container	4	
Q235	Do you clean the containers regularly? of water storage?	Yes	1	
		No	0>	If 0! Q237
Q236	At what frequency?	Every week	1	
		Every 2-3 days	2	
		Every day	3	
		Before filling each container	4	
Q237	How do you use stored water? You?	Ladle - utensil	1	
		Valve on container	2	
		Water poured directly	3	
Q238	When do you refill your storage containers?	Once they are empty	1	
		As we go along	0	
Q239	How long on average is the water- Is it stored?			
Q240	Wash your hands before you use any equipment. water handling?	Yes - with soap	1	
		Yes - without soap	2	
		No	3	
Q241	How does the water you use look? consume?	Cloudy - turbid	1	
		Red	2	
		Verdatre	3	
		Clear - colourless	4	
		Presence of foreign bodies	5	
Q242	How does your water taste?	Neutral - no taste	1	
		Fade	2	
		Dirty	3	
		Bad taste	4	
Q243	Does your water look clean to you? consumption?	Yes	1	
		No	0	
Q244	Do you think that water can be a source of certain diseases?	Yes	1	
		No	2	
		Don't know	3	
Q245	Are you experiencing any problems related to the water consumption?	Yes	1	
		No	0>	If 0! Q247
Q246	What kind of problem do you have?	Dermatological disease	1	
		Urinary-renal disease	2	
		Intestinal disease - dysentery-diarrhea	3	
		Cholera	4	
		Typhoid	5	
		Parasite	6	
Q247	Do you think that water treatment could prevent disease?	Yes	1	
		No	0	
Q248	Would you be willing to use tablets to improve water quality?	Yes	1>	If 1! Q301
		No	0	
Q249	Why?			

II-Hygiene and Sanitation

N°	Questions	Answers		Instruction
Q301	How often do you clean your yard?	Every day Every second day Every third day or more Never	1 2 3 4	
Q302	How do you dispose of waste water?	Concession	1	
		Lost wells	2	
		Street	3	
Q303	How do you dispose of rubbish and animal waste?	Incineration	1	
		Rubbish heap Lost well Street	2 3 4	
Q304	Do you think that waste water and rubbish left in the wind can cause disease?	Yes	1	
		No	0>	If 0! Q306
Q305	Which ones?			
Q306	Where do you relieve yourself?			
Q307	Are there latrines in your yard?	Yes	1	
		No	0>	If 0! Q311
Q308	What type of latrine?	Traditional	1	
		Modern	0	
Q309	How is your latrine emptied?	Manual	1	
		Mechanics	0	
Q310	What do you do with the excreta?			

Q311	Do you wash your hands after relieving yourself?	Yes to soap Yes without soap No	1 2 3	
Q312	How far is the rubbish dump from your water source during the rainy season?			
Q313	How far is the rubbish dump from your water source in the dry season?			
Q314	Do you wash your hands with soap before each meal?	Yes No	1 0	

Annex 2: Questionnaire for water point managers

Identification

N°	Questions and Filters		Codes
001	Questionnaire number		L_I_J_
002	Region		I_I
003	Province		I_I
004	Municipality		I_I_I
005	Village		I_I I
006	Neighbourhood		
007	Location of the water point		X: Y :

* Result Codes
1=Completed questionnaire
2=Refusal
3=Different
4=Absent survey
5 = Partially filled

I. Socio-demographic characteristics of the respondent(e)

N°	Questions	Answers		Instruction
Q101	Name of the survey			
Q102	Gender	Female Male	0 1	
Q103	Age	Years gone by / / / years		
Q104	Status of the survey			
Q105	Did you go to school?	No Yes	0> 1	If 0! Q107
Q106	What level of education did you attend?	Primary Secondary Higher	1 2 3	
Q107	Have you attend a centre of literacy?	No Yes	0 1	
Q108	What level have you reached?	Alpha Initial Basic Complementary Alpha Specialised Technical Training	1 2 3	

II-Water point

N°	Questions	Answers		Instruction
Q201	What is the type of source?	Surface water Well Drilling Water system	1 2> 3> 4>	If 2! Q203 If 2! Q205 If 4! Q204
Q202	What is the type of surface water?	Watercourse 1 Lake 2 Mare 3 Dam 4 ᴶ		If 1, 2, 3, 4! Q207

N°	Questions	Answers			Instruction
Q203	What is the type of well?		Yes	No	
		Traditional	1	0	
		Modern	1	0	
		Protects	1	0	
		Unprotected	1	0	
Q204	What is the type of water system?	Fountain bollard Individual connection	1 0		
Q205	Is there any ancillary infrastructure?	Yes No	1 0		
Q206	Which ones?				
Q207	How many people use this water source?				
Q208	Is there anyone in the community who does not have access to a water point?	Yes No Don't know	1 2}		If 2, 3! Q211
Q209	How many people do not have access to water?				
Q210	Why doesn't everyone have access to the improved water system?	The system does not provide enough water It is too expensive The system is too far away Queues are too long The system does not reach all community members Don't know	1 2 3 4 5 6		
Q211	In what year was the water point built?				
Q212	Has the water point been rehabilitated yet?	Yes No Don't know	1 2}		If 2, 3! Q214
Q213	In what year?				
Q214	Is the water point part of a project supported by	Yes	1		If 0! Q217

N°	Questions	Answers		Instruction
	an organisation?	No	0>	
Q215	Which organisation supports this project?			
Q216	What role did the organisation play?	Technical advice	1	
		Equipment	2	
			3	
		Software/Training/Knowledge support Financial	4	
		support		
Q217	Is water treatment done at the water point?	Yes	1	
		No	0>	If 0! Q220
Q218	What water treatment is done?	Chlorination	1	
		Don't know	0	
Q219	By whom is it made?			
N°	Questions	Answers		Instruction
Q220	Why?			
Q221	Does the water source provide enough water for the community every day of the year?	Yes	1>	If 1! Q225
		No	0	
Q222	Why?	Long periods of unavailability Irregular supply	1	
		Low water pressure Political reasons Payment	2	
			3	
		problems Disconnected tap Rationing	4	
			5	
		Seasonal stables Don't know	6	
			7>	
			8>	If 7! Q224
			9	If 8! Q223
Q223	In which months do these shortages occur?		Yes	No
		January	1	0
		February	1	0
		March	1	0
		April	1	0
		May	1	0
		June	1	0
		July	1	0
		August	1	0
		September	1	0
		October	1	0
		November	1	0
		December	1	0
Q224	How many hours a day is water available?			
Q225	Has the system failed for more than one day in the last 30 days (except for routine maintenance)?	Yes	1	
		No Don't know	2 3 }	If 2, 3! Q228
Q226	How many days did it take for it to be repaired?			
Q227	Why was the system down for more than a day?	Lack of money Rationing No plumber/person	1	
		who knows how to repair Political reasons Parts	2	
			3	
		not available Don't know	4	
			5	
			6	
			7	
Q228	Have there been any major repairs/changes to the water point/system in recent years?	Yes	1	
		No	2 }	If 2, 3! Q231
		Don't know		
Q229	What has been repaired/replaced/removed?			
Q230	How much did the repair/improvement cost?			
Q231	Are there any current problems with the system that require intervention?	Yes	1	
		No	0>	If 0! Q233
N°	Questions	Answers		Instruction
Q232	What are the existing problems?	Broken parts	1	
		Poor water quality	2	
		Low water flow	3	
Q233	Is water from the spring available on the day of the visit?	Yes	1>	If 1! Q235
		No	0	
Q234	Why is water not available on this day?	System is broken System is under maintenance or	1	
		repair No community member is available to	2	
			3	
		authorise water collection Rationing Seasonal	4	
		shortages	5	
Q235	Is there a tariff or user fee for water from the water point?	Yes	1	
		No	0>	If 0! Q239
Q236	How often are user charges levied?	Per unit (canister - barrel...) Per day Per week		
		Per month		
		By season		
		By year		
		Package plus unit price		

N°	Questions	Answers		Instruction
Q237	What is the rate/amount of the user fee?			
Q238	Who is in charge of the payment?	The woman of the house The man of the house Anyone from the household	1 2 3	
Q239	Does the community have spare parts on hand for the water point?	Yes No **Don't know**	1> 2 3>	If 1, 3! Q241
Q240	Are spare parts available from the community?	Yes No Don't know	1 2 3	
Q241	Who manages the water point?	Water Committee/Office Individual owner of the system Owner's employee Private company (company - cooperative...) Government NGO Church Group No designated manager Don't know	1 2 3 4 5 6 7 8 9	
Q242	If there were problems with the water point that could not be handled by the community, who would you contact?			
N°	Questions	Answers		Instruction
Q243	Has the water point/system been able to support new users since it was built/enabled?	Yes No Don't know		
Q244	What equipment is used to draw water?	Hand pump Bucket and rope Bucket Can	1 2 3 4 5	
Q245	Where is the water collection container stored when not in use?	Remains in the well/workplace Outside - on the ground In a public place In someone's home	1 2 3 4	
Q247	Is the material used for water collection cleaned before being immersed in water?	Yes No	1 2	
Q248	OBSERVATION: Colour of the water?	Rougeatre Verdatre Brown Yellow Colourless	1 2 3 4 5	
Q249	OBSERVATION: Water turbidity?	Very cloudy water Cloudy water Fairly clear water Colourless water		
Q250	OBSERVATION: Presence of foreign bodies?	Yes No	1 0	
Q251	OBSERVE: Is the environment around the water point/system clean, maintained?	Yes No	1 0>	If 0! Q253
Q252	Who maintains it?	Person Committee- maintenance service/hygiene Maintenance employee Everyone	1 2 3 4	
Q253	What are the other uses around the waterhole?			

Observations :

Annex 3: The interview guides
The opinion leader interview guide
I .ocalite :

Name First name(s)

Status:

Problems of access to drinking water.

The actions undertaken by the town hall to improve the situation.

the existing water infrastructure.

Water infrastructure project in your community

Likely dates for infrastructure development.

The interview guide for health workers
Name of the health centre :

Name :first name(s) :

Grade :
Waterborne diseases recorded in your facility
Statistics
Most affected age group
Main causes of these waterborne diseases
Patient counselling
Patient compliance with your advice
Strategies to improve the health situation

Annex 4: Water point collection sheet

REGION	PROVINCE	COMMUNE	VILLAGE	QUARTIER	COORDONNEES			FONCTION-NALITE	TYPE	DESCRIP-TION
					X	Y	Z			

FUNCTIONALITY: F=FUNCTIONAL, NF=NON FUNCTIONAL
TYPE : PT=TRADITIONAL WELLS, PD=LARGE DIAMETER WELLS, F=DRILLING

Printed by Books on Demand GmbH, Norderstedt / Germany